D0006681

SURVIVAL SKILLS HANDBOOK

VOLUME 1

CAMPING

•

MAPS & NAVIGATION

•

KNOTS

•

DANGERS & EMERGENCIES

CONTENTS

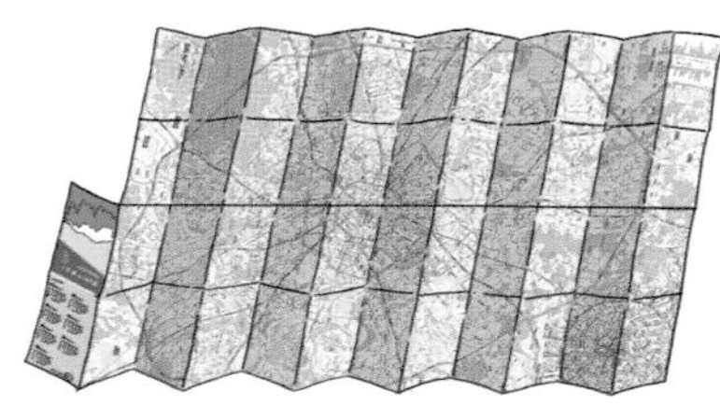

KNOTS

DANGERS AND EMERGENCIES

INTRODUCTION

We live on the most amazing planet, and should seize every chance we get to explore it. From the hottest desert to the highest mountain, I have been on lots of exciting adventures around the world, and seen some phenomenal parts of our planet. When you're out exploring, it's important to be prepared and to stay safe in the wild. Make sure you know about the area you are planning to go to beforehand—what is the weather like? What dangers are you likely to face? Once you know as much as you can, you will need to decide what equipment to take. It is very important to plan for emergency situations. When you're out in the wild, good preparation can be the difference between life and death. It is also vital that you are respectful of the world around us, and leave everything just as you found it. Exploring the world is a great privilege, and it is our responsibility to take care of our beautiful planet so we can continue adventuring for many years to come.

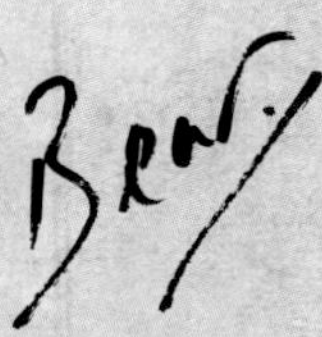

CAMPING

Camping in the great outdoors can be one of the most rewarding experiences—as long as you are fully prepared for a night under the stars. Once you know how to build a campfire, safely store food, and build your shelter you can embark on great adventures!

IN THIS SECTION:

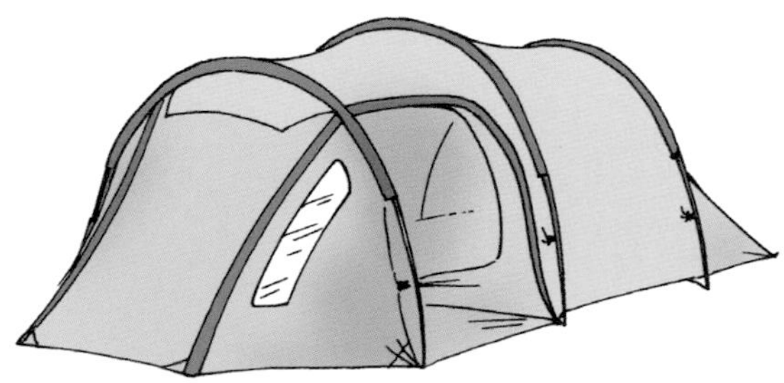

GOING CAMPING!

Camping with friends and family can be a mini-adventure! A good campsite will allow you to enjoy your surroundings and should be kind to the environment. You should also be able to relax there safely. Take the time to fully prepare for your trip and you'll make great memories in the outdoors.

Making camp

Camp at existing campsites when possible, and remember that good campsites are found, not made.

BEAR SAYS

Camping is a great way to enjoy the outdoors and experience nature at its best!

Places to avoid

Some locations are not good for camping—stay away at all costs!

Flood risk

Washouts, gullies, and floodplains can be deadly when it rains.

Under a tree

Even healthy-looking branches may drop without warning.

Cliff base

Don't camp below a cliff or a steep rocky slope in case loose rocks fall.

Avalanche risk

Stay away from steep slopes during or after heavy snowfall.

TENTS

A tent is your home away from home while camping. It shelters you from wind, rain, cold temperatures, and blazing sun. There are lots of different types to choose from.

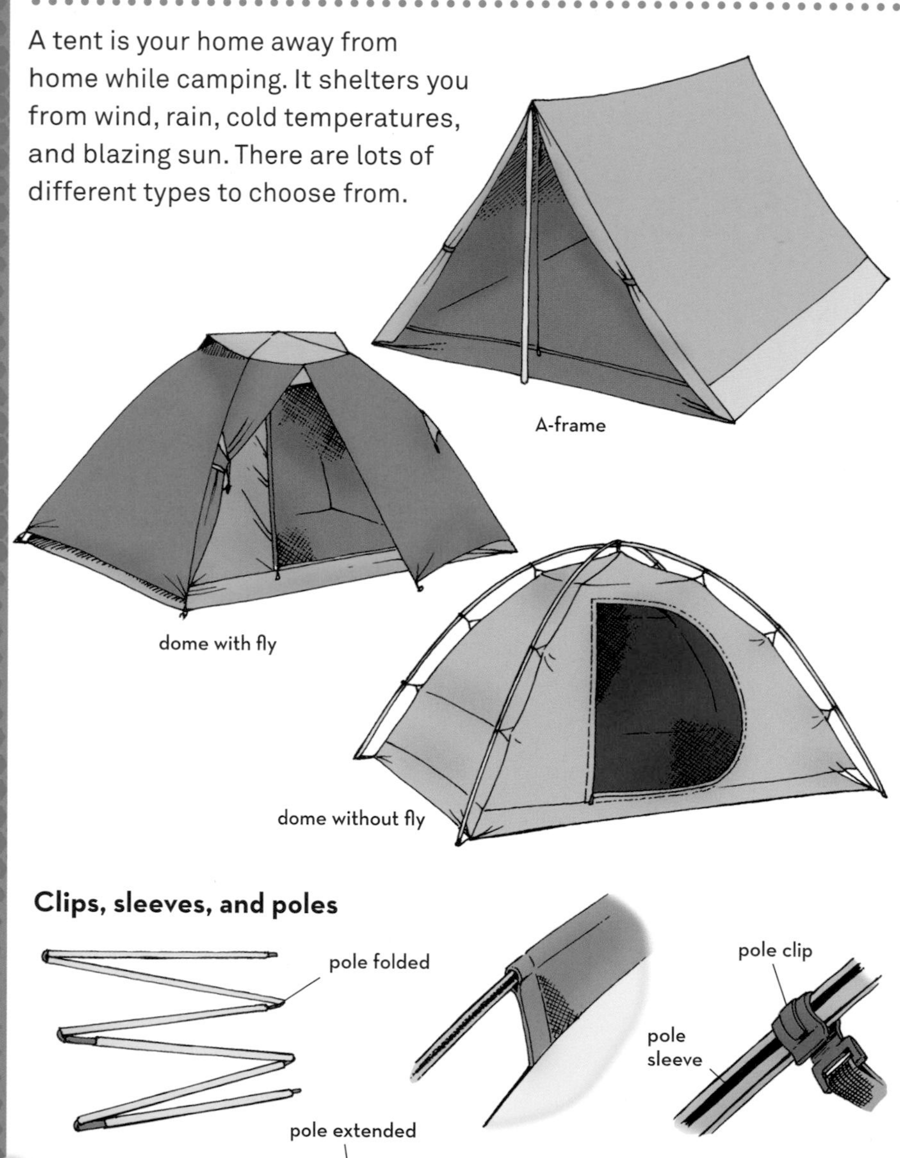

Clips, sleeves, and poles

geodesic
two-hoop
three-hoop
family
self-erecting

Pegs

Very few tents will stay up by themselves, and only then in good weather. They need guy ropes (ropes under tension) and secure attachments to the ground. That's where pegs come in.

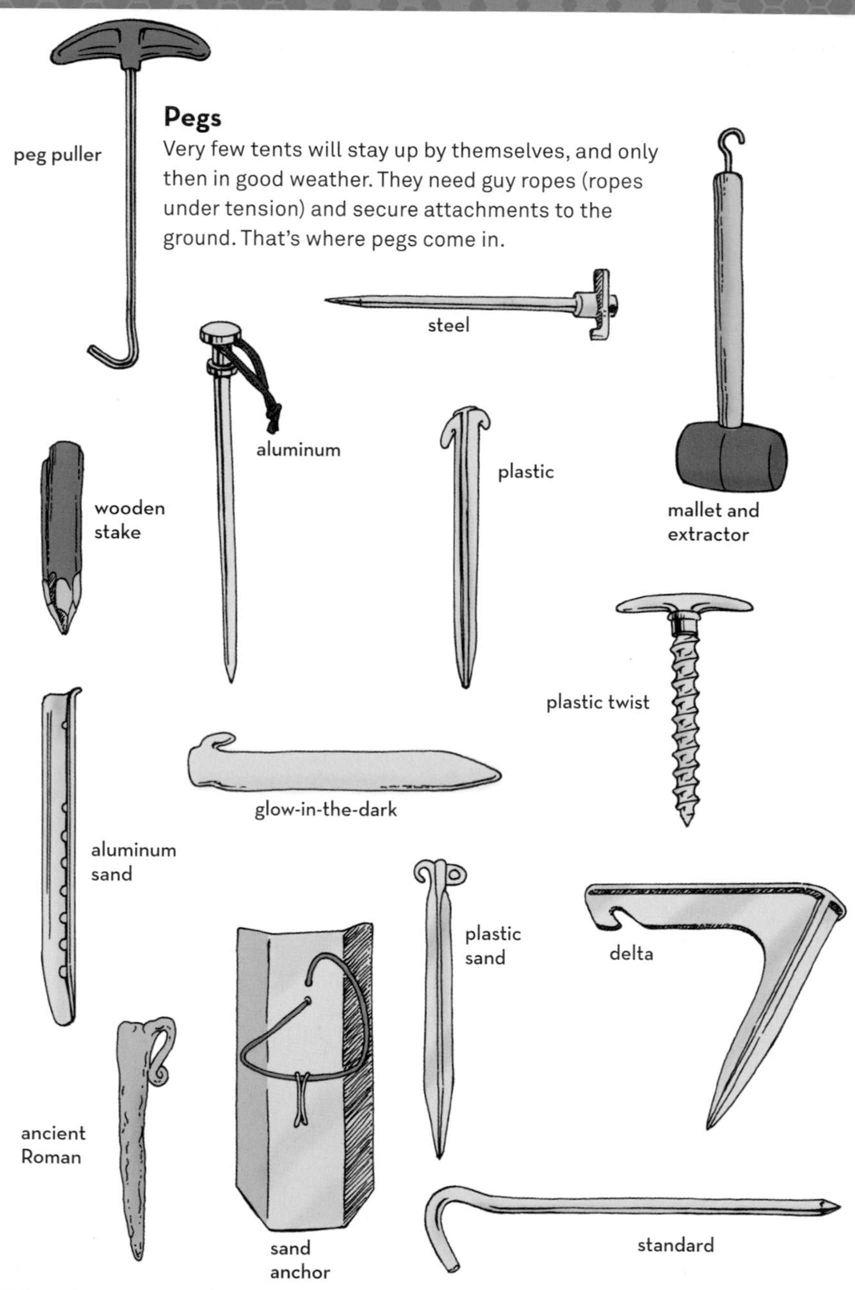

Placing pegs

It is important to make sure your tent is stable and stays put so that you have proper shelter.

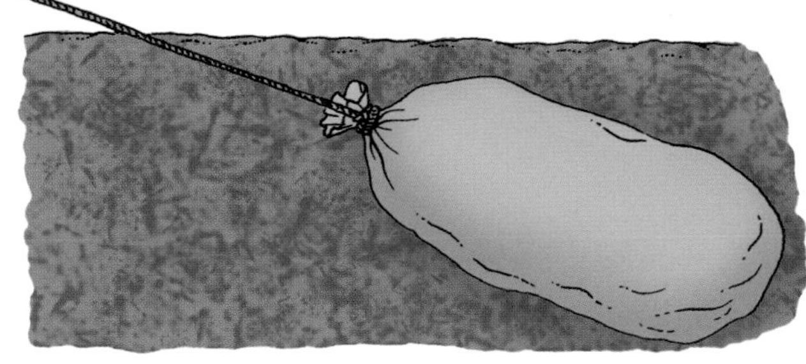

Mountain favorite

A buried sleeping bag case filled with snow makes a good anchor.

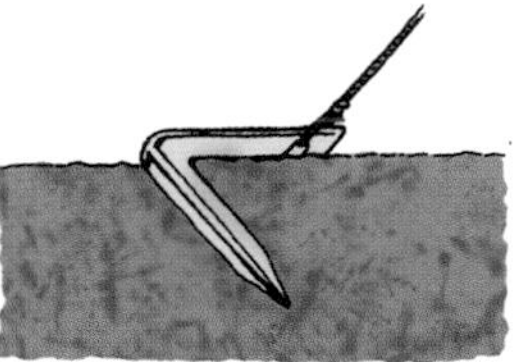

Standard

The rope is at 90 degrees to the peg.

Super stable

When stability is key, use two pegs for extra security.

Delta

These strong pegs keep a very secure hold.

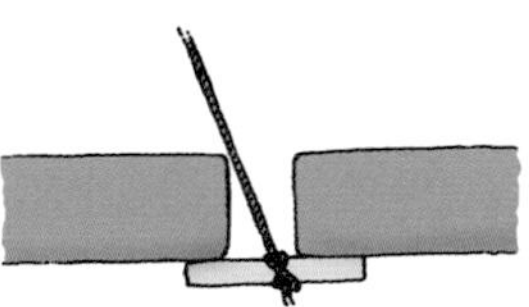

Back-up

Use a heavy rock to secure a peg if it won't go far into the ground.

Rocks

Piles of rocks can be used to anchor your lines.

Ice

A tent peg can be placed in a hole in the ice.

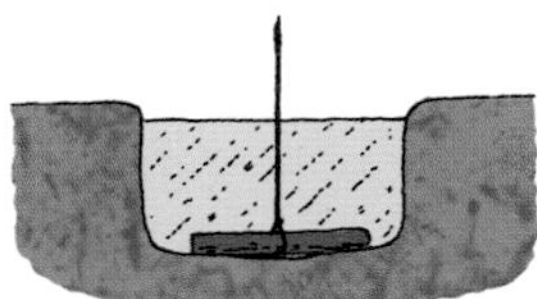

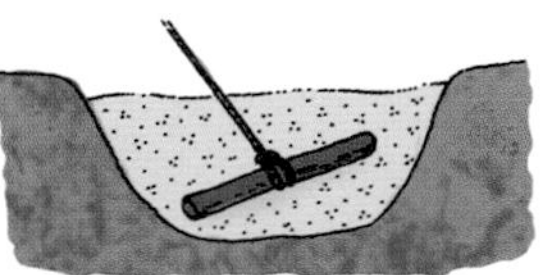

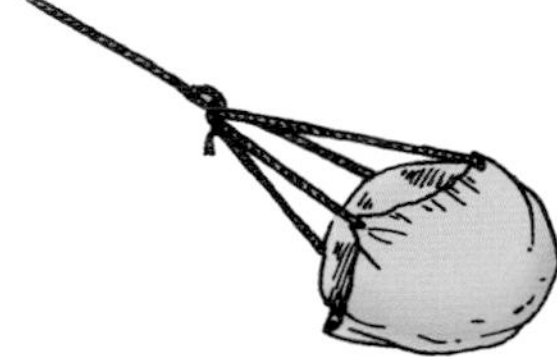

Desert sand

A peg buried in sand makes a good anchor.

Buried in ice

A peg can be frozen in ice to secure its hold.

Parachute

Place heavy objects in a parachute anchor, then bury.

What to do if your tent leaks

If you expect wet weather or a downpour looms, these simple tips can save the day.

Water trench

A bit of digging can help avoid a flood in your tent.

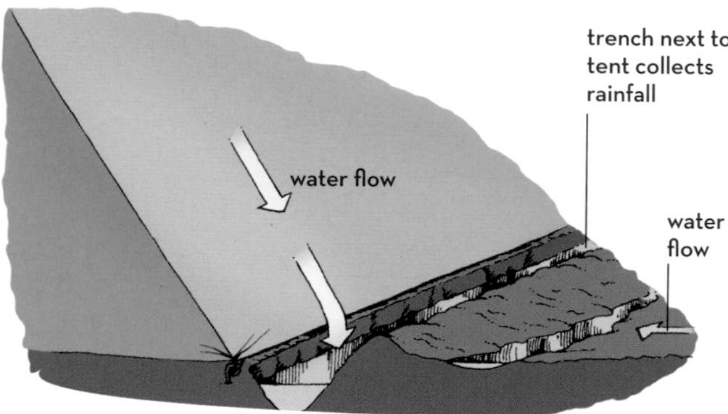

Inside drip cord

This short-term fix allows leaking water to drip into a bowl rather than your sleeping bag.

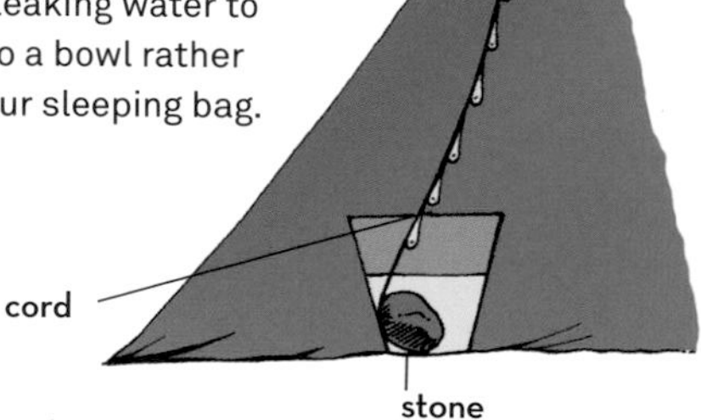

Outside cord

This diverts water from reaching the anchor point of your tent or tarp.

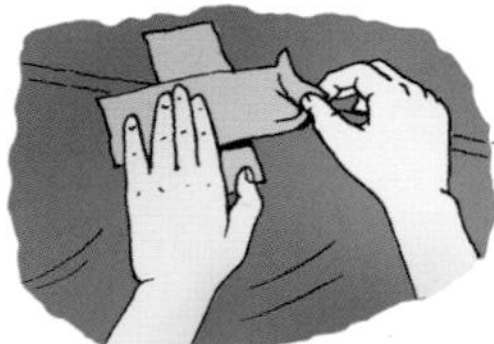

Duct tape

Apply to a hole or split seam.

Hole patch

Patch kits will securely seal a leaking hole.

Seam sealer

Use a sealer for leaks along the seam.

OTHER TYPES OF SHELTER

From tarpaulin to igloos, there are many other types of shelter that could be a better option than a tent.

The art of tarp

These are some ways tarpaulin can be made into a shelter.

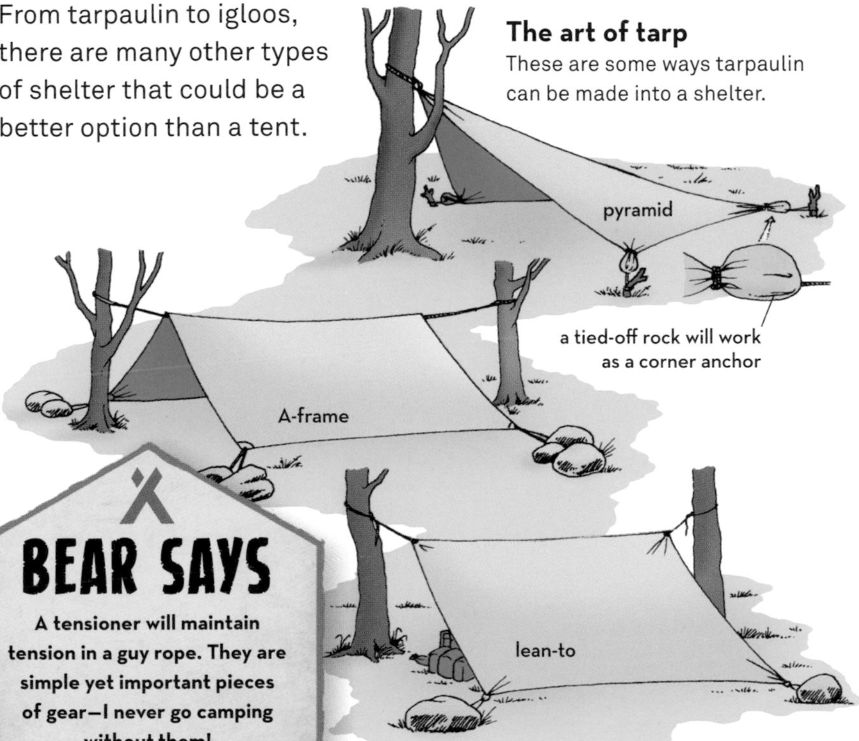

pyramid

a tied-off rock will work as a corner anchor

A-frame

lean-to

BEAR SAYS

A tensioner will maintain tension in a guy rope. They are simple yet important pieces of gear—I never go camping without them!

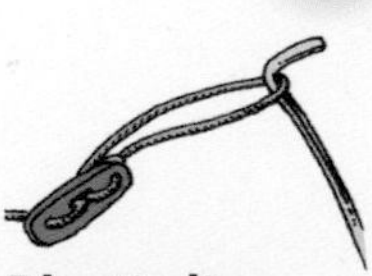

Plastic line

A small tug on this tensioner will tighten the guy ropes.

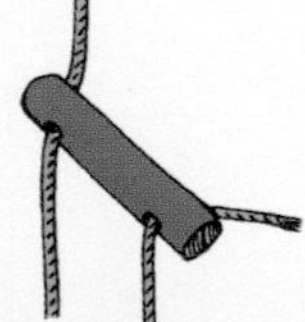

Timber line

A piece of timber with two holes will also hold guy ropes in place.

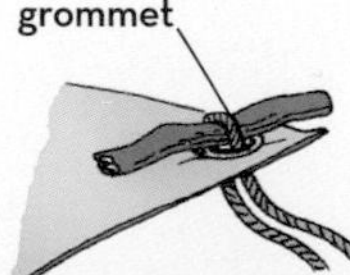

Stick anchor

This simple method will save wear on your tarp corner grommets.

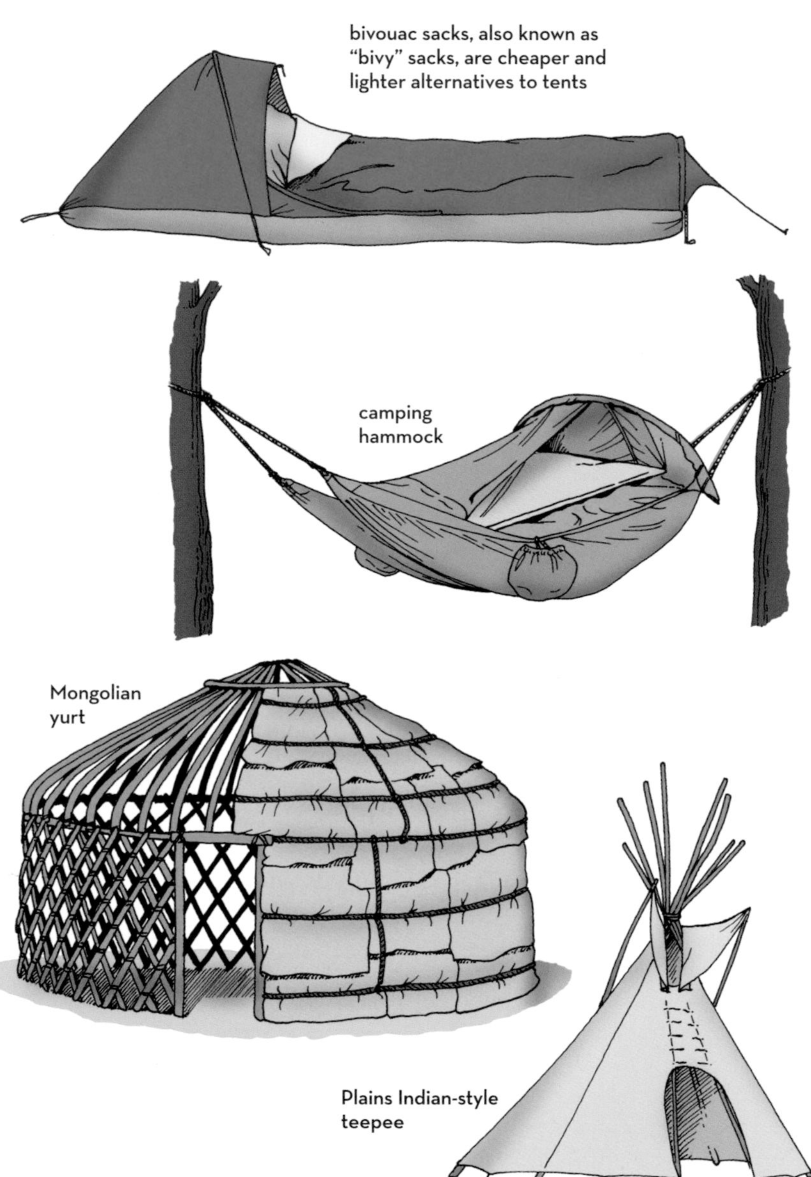
bivouac sacks, also known as
"bivy" sacks, are cheaper and
lighter alternatives to tents
camping
hammock
Mongolian
yurt
Plains Indian-style
teepee

Snow cave

With experience and a snow shovel, you and a friend can build a comfortable snow cave in a couple of hours. They can be a lifesaver, but you have to be careful of carbon monoxide gas poisoning. Always make sure there's a good flow of fresh air.

Igloo

This Inuit invention is an option for shelter in cold conditions.

BEAR SAYS

Snow is a great insulator in freezing conditions. Once you have made your shelter, compress snow into blocks to seal up the entrance.

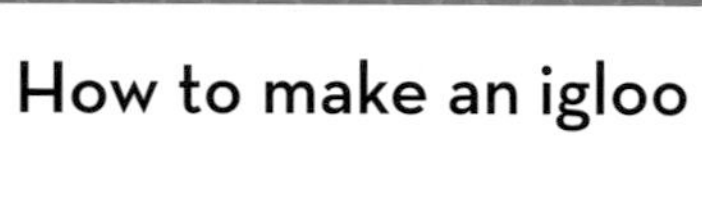

How to make an igloo

1 Mark a circle in the snow about 7 ft. across.

2 Tamp down the snow inside the circle until you have a solid surface.

1. To make a snow block, first cut two parallel lines.

snow saw

2. Then make a horizontal cut.

3. Last, make a vertical cut.

3 Using a snow saw, cut blocks of hard, compacted snow. Hard snow can usually be found below soft snow.

5 Cut a ramp in the snow blocks halfway around the circle.

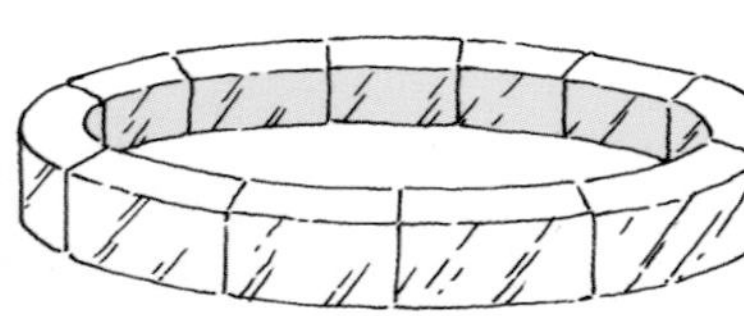

4 Arrange your first blocks in a circle.

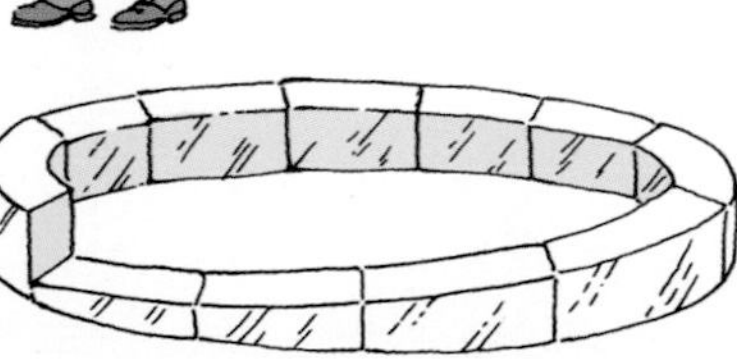

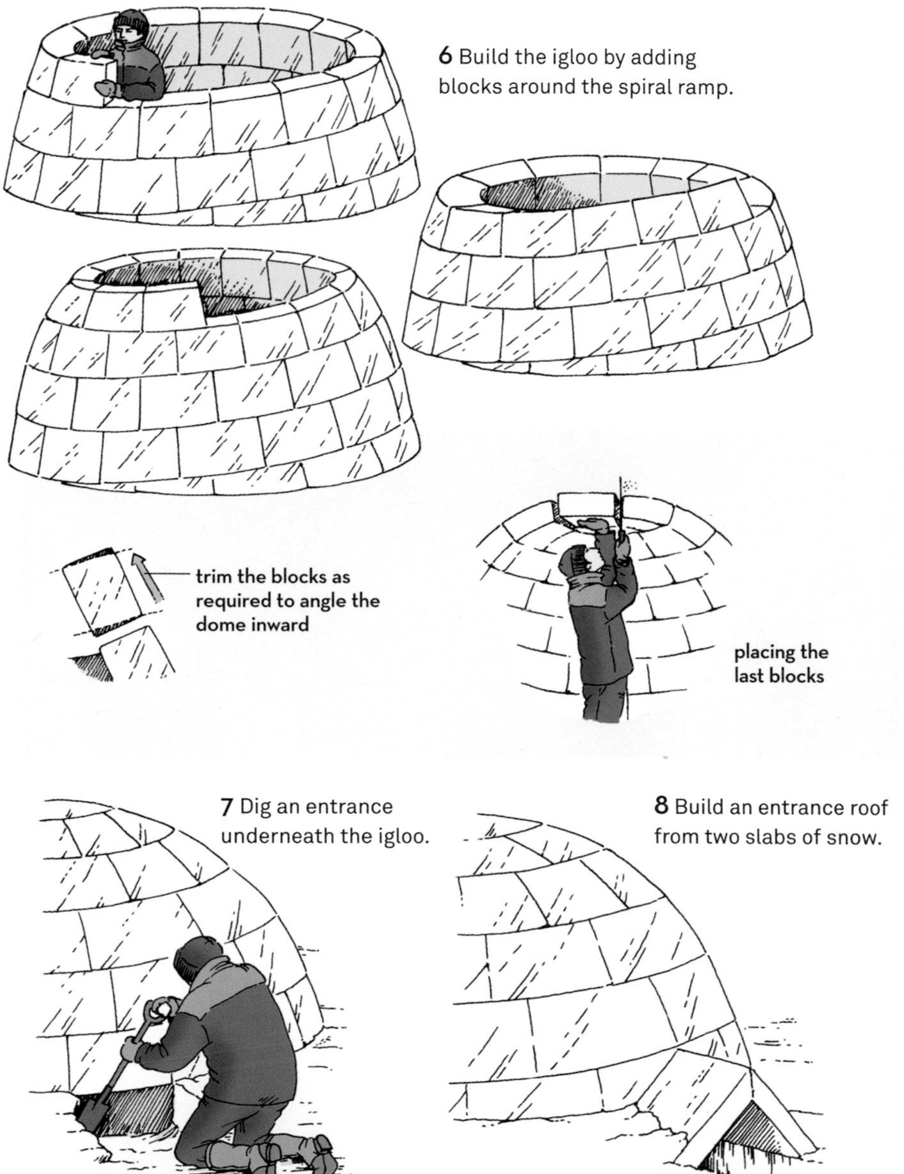
6 Build the igloo by adding blocks around the spiral ramp.
trim the blocks as required to angle the dome inward
placing the last blocks
7 Dig an entrance underneath the igloo.
8 Build an entrance roof from two slabs of snow.

How to make a quinzee

A quinzee is a large pile of snow that has been hollowed out.

1 Put backpacks and any other bulky gear together and cover with a tarp.

2 Pile up a good-sized amount of snow over your backpacks.

3 Pack down the snow and wait a couple of hours while it "sinters" (this is when the snow crystals bind to each other).

4 Stick even-lengthed sticks all around the snow pile.

5 Excavate the snow. The other ends of the sticks will guide you and keep you from digging through.

6 Insulate the base with tarps or sleeping pads, and make yourself at home.

SLEEPING SOFTLY

A camp mattress makes sleeping more comfortable, and also keeps you warmer than if you were to sleep directly on the ground.

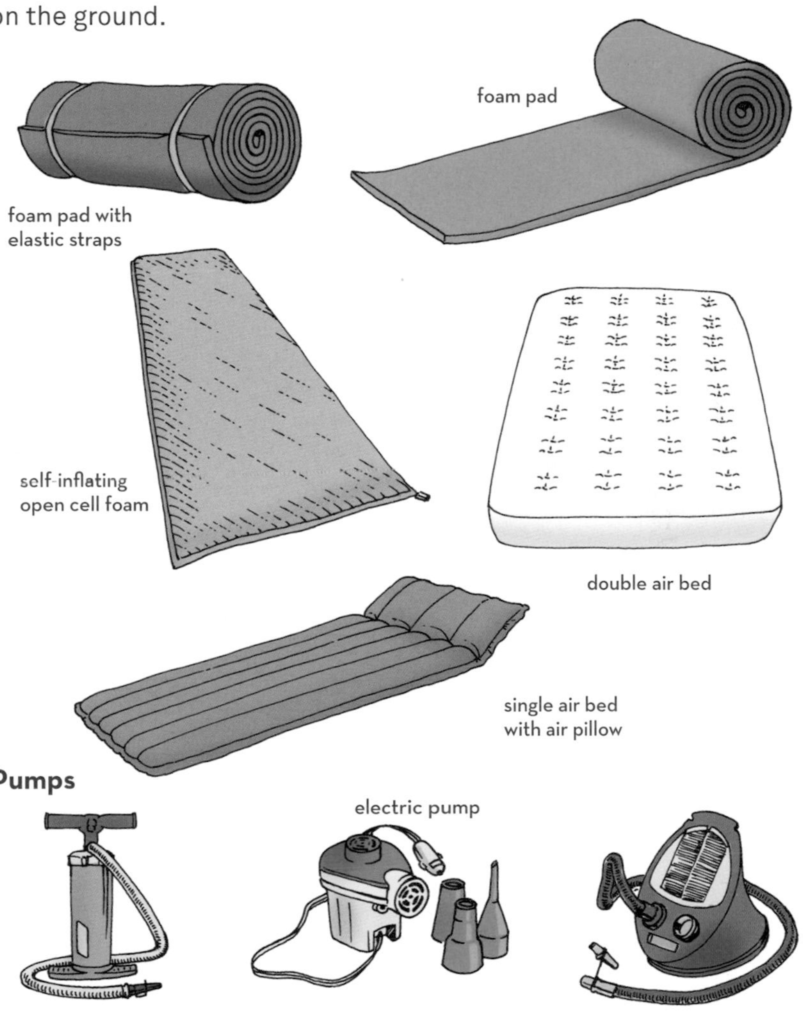

Pumps

hand pump

electric pump

foot pump (uses bellows)

SLEEPING BAGS

A good-quality sleeping bag is essential for a good night's sleep in the outdoors.

BEAR SAYS

Sleeping bags are such an important piece of gear! You'll want to make sure you have the right type depending on the temperature at night.

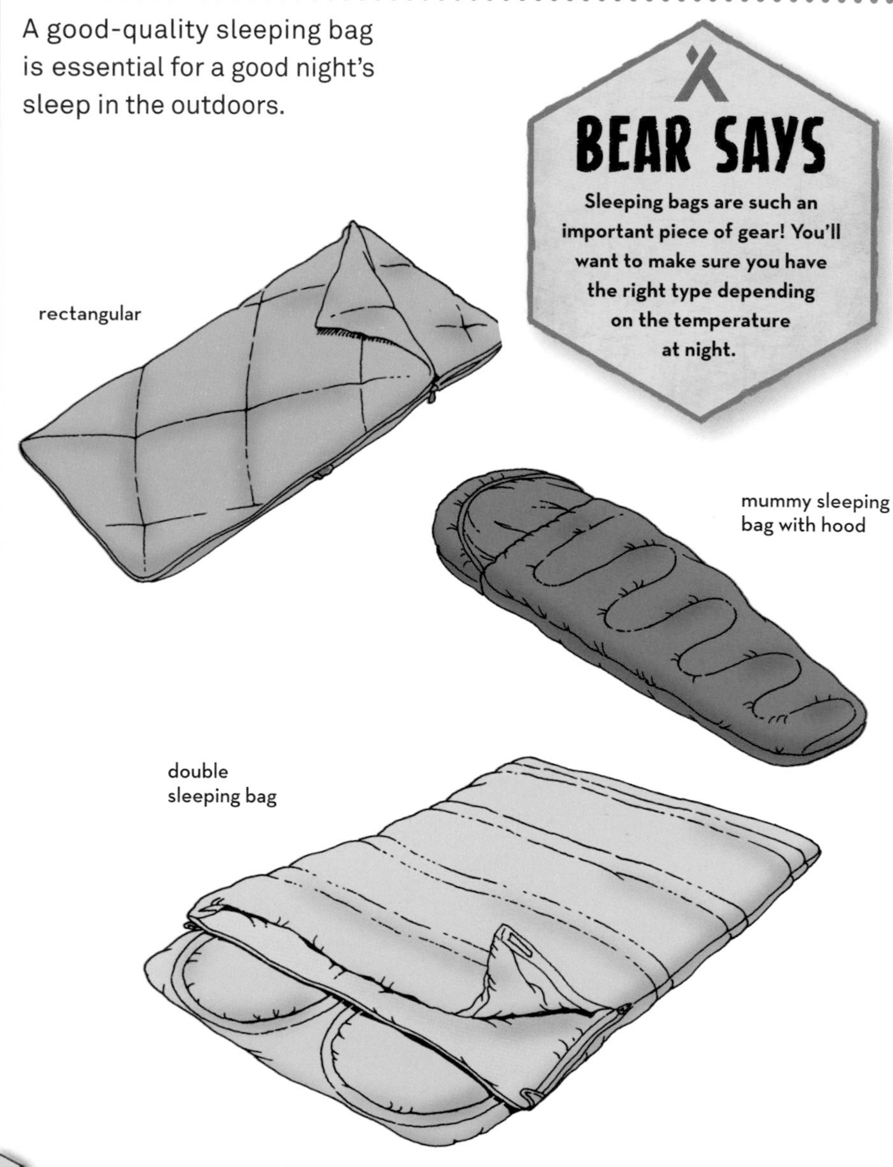

compression sack

stuff sack

Sleeping bag fillings

A sleeping bag is made up of a lining and an outer shell. In between the two are different types of filling. The way in which the filling is stitched together affects how warm you stay inside it. Down bags are often built with pockets called baffles that stop the down from bunching.

sewn through—outer shell is stitched together

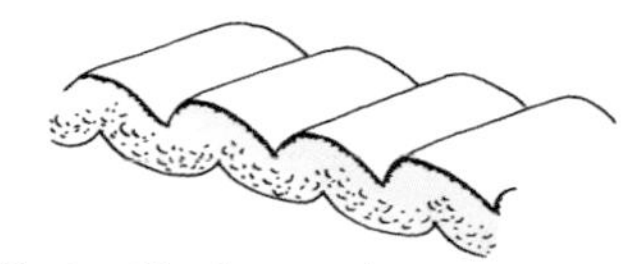

offset quilt—staggered double layers

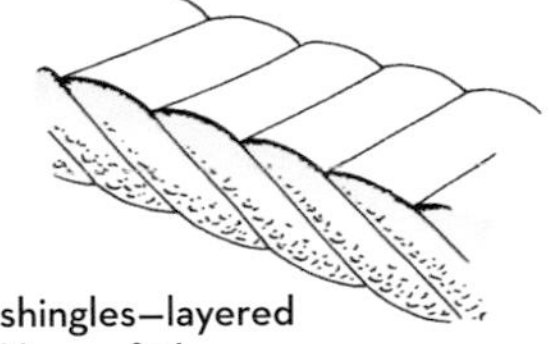

shingles—layered like roof tiles

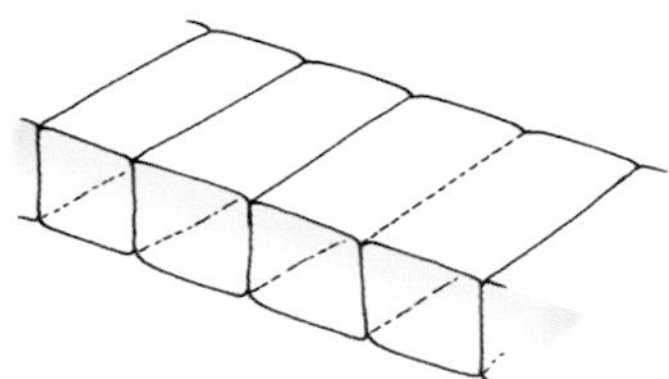

box-shaped baffle

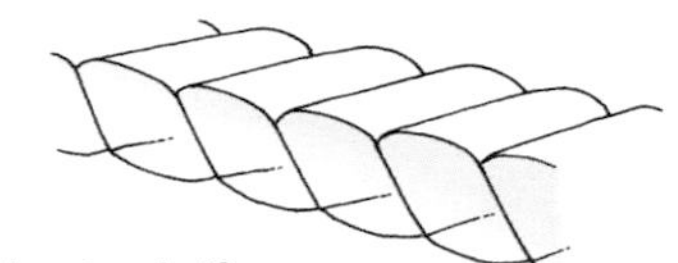

slant box baffle

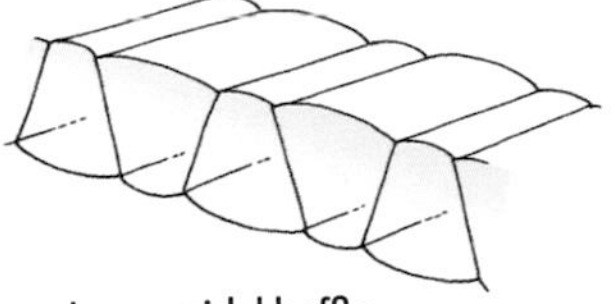

trapezoidal baffle

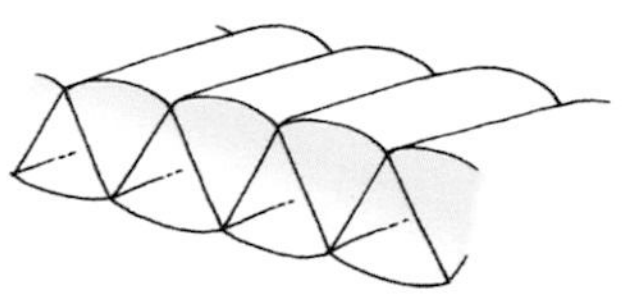

V-tube baffle

KNIVES

A good sharp knife is a camping essential but should only be used by an adult with caution and care.

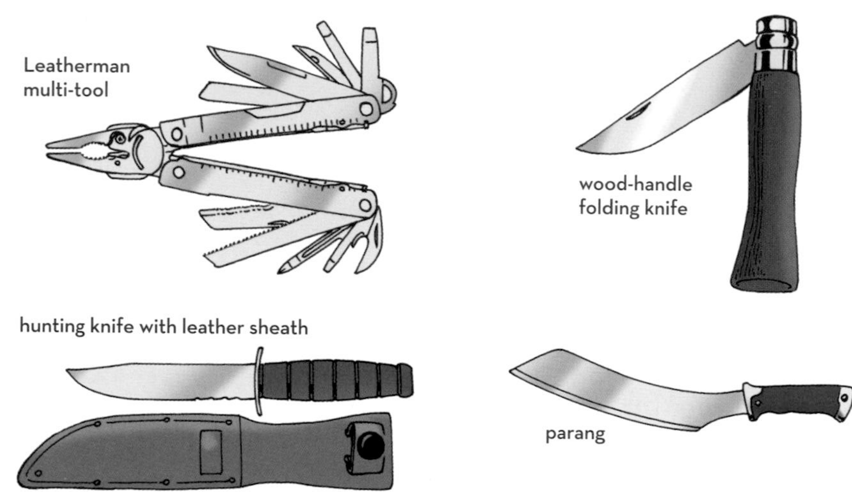

Survival knife

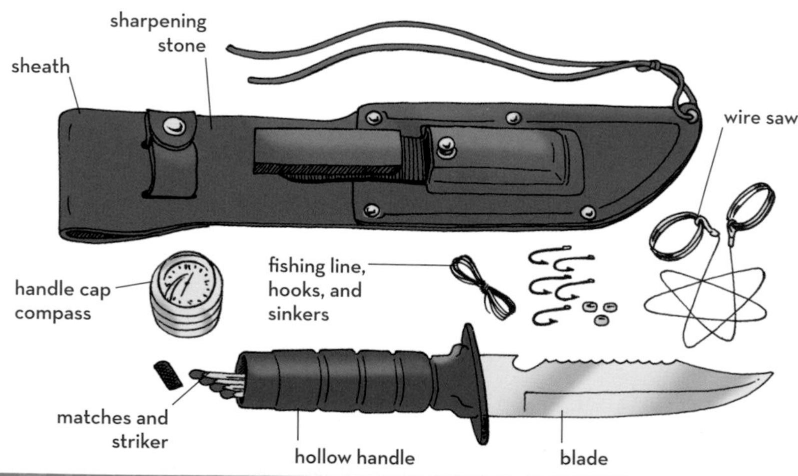

BEAR SAYS

Knives can have fixed or folding blades, and often come with other tools too.

Parts of a hunting knife

Knife sharpening

To sharpen a knife, the blade is often passed over a hard, rough surface.

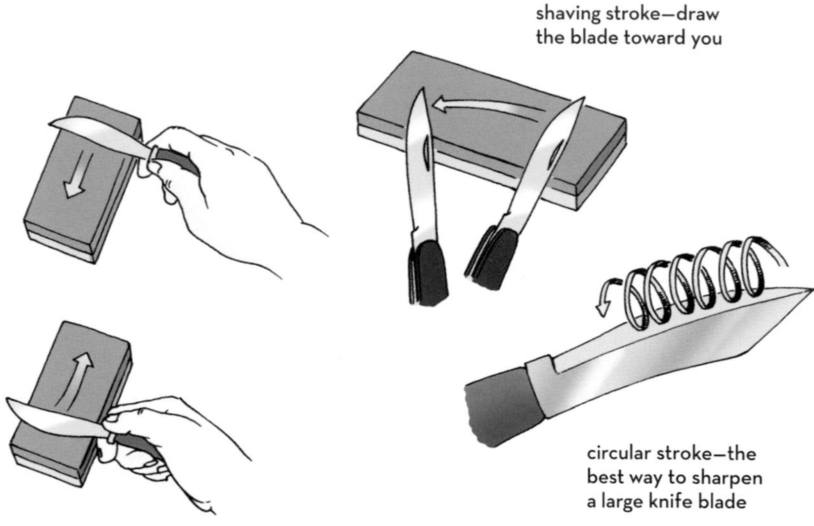

Useful tools

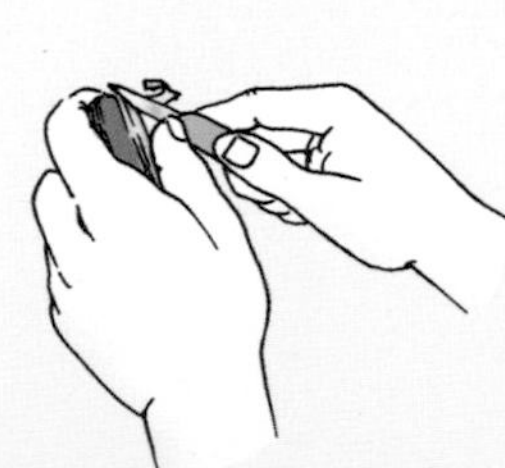

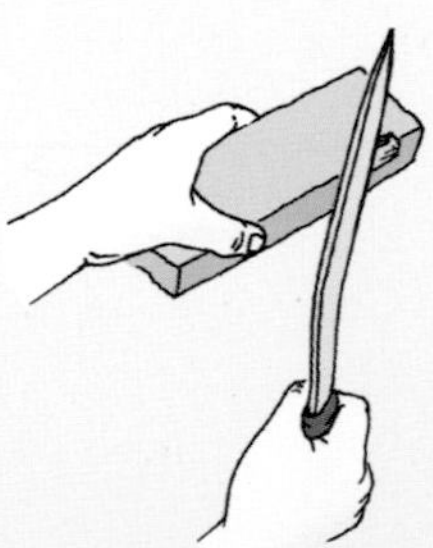

Chop

Only large blades should be used for chopping.

Whittle

Use a small knife for fine carving.

Carve

To carve, make shallow cuts along the grain.

honing oil

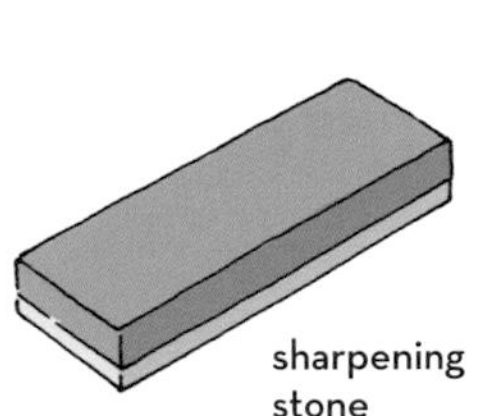

sharpening stone

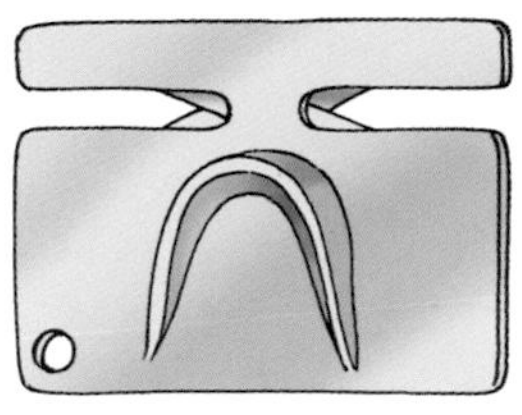

sharpening tool

sharpening steel

BEAR SAYS

Knives are very useful pieces of equipment and they require a lot of special care.

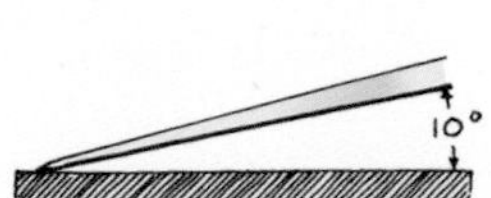

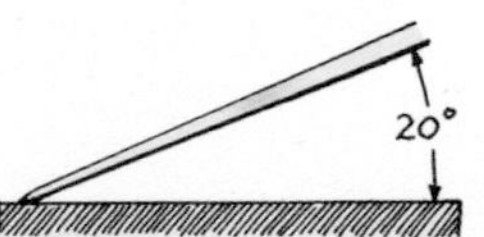

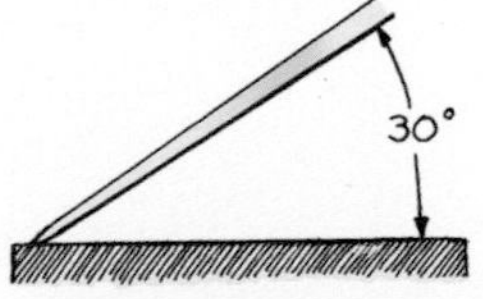

10 degrees

For light duty and fine work such as filleting and shaving. The edge will blunt fairly quickly.

20 degrees

A good angle for everyday use. To approximate it, imagine half of 90 degrees, then half of that.

30 degrees

A somewhat blunt yet long-lasting edge for heavy-duty work, such as chopping wood.

TOILETS AND SHOWERS

For a healthy environment and a healthy you, make sure you maintain hygiene by washing and showering. Depending on your needs, there are different types of toilets that can be built outdoors. Be sure to keep toilets away from water sources and trails.

Cat hole

For one-time, personal use.

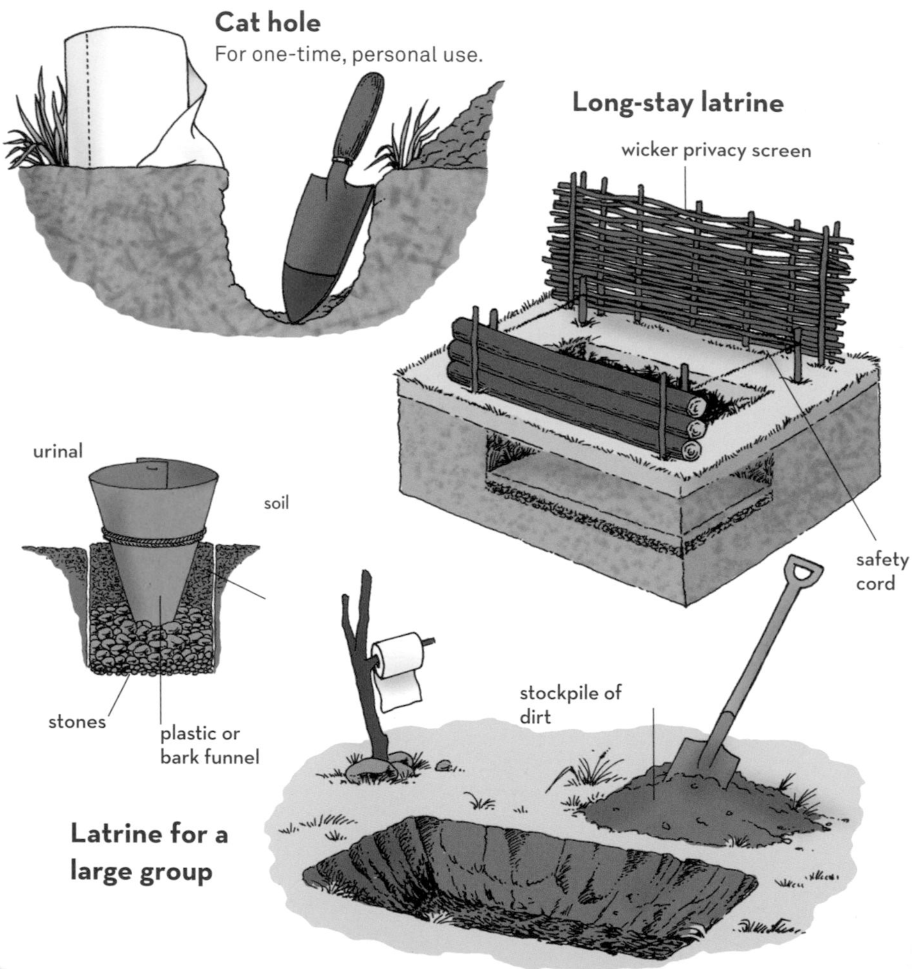

Keeping clean

soap

hand wash

antibacterial wipes

Using a solar shower

1 Fill the shower bag with water and lay it in the sun. On a cool or overcast day, do this in the morning so that there is warm water in the afternoon.

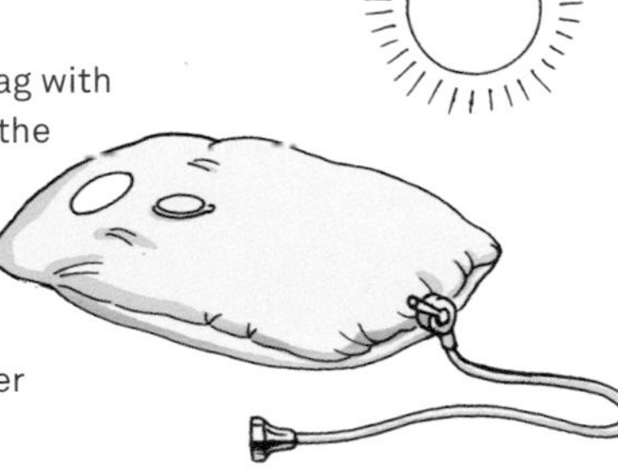

2 Hang the bag from a tree. The bag will be heavy, so pick a branch that is strong and healthy.

3 Check that the water is at a safe temperature before starting your shower.

BEAR SAYS

Staying clean may not always be an obvious part of survival —but without it hygiene will suffer, making infections more likely.

you could hang a towel as a curtain for privacy

FOOD CACHES

You don't want to share your precious food supplies with the local wildlife. In some places hungry bears may look for a meal.

Bear caches

Traditional cache

These mini log cabins are raised up high off the ground. They are still used in North American woods.

Campground cache

These permanent bear-proof cabinets are a common sight in campgrounds where bears might visit.

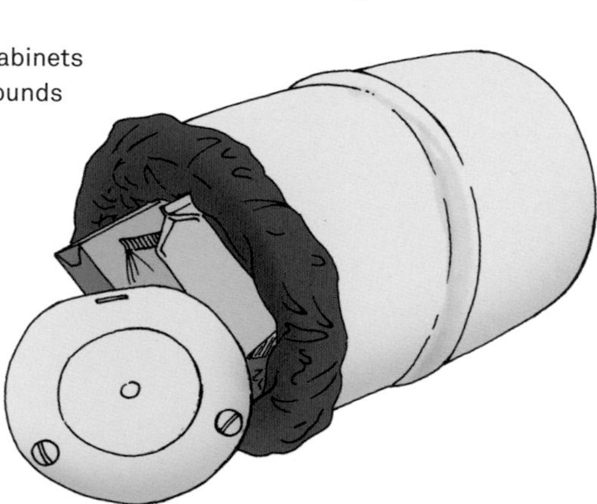

Bear can

These tough canisters can hold about a week's worth of food for the average hiker.

Setting up a throwline bear cache

1 Find two trees about 15 ft. apart. Throw the rope up and over a branch.

2 Tie the rope to the tree trunk then throw the rope over a branch on the second tree.

3 Then, hoist it up until it's at least 12 ft. above the ground.

4 Tie the other end of the line to the trunk of the second tree. Your food is now safe.

BEAR SAYS

Bears have an incredible sense of smell, and are also able to recognize food containers.

NUTRITION

Eating well is the key to good health. This is especially true if you are getting active outdoors. Make sure you get a balanced intake of water, various food groups, vitamins, and minerals.

BEAR SAYS

On days when you are active you will need more food and water than usual. Always stay hydrated—especially in hot weather.

Water

The most vital substance in our bodies is water. It makes up more than half of a person's bodyweight, and fulfills such important roles that even a few days without it can be fatal.

Micronutrients

Micronutrients are essential vitamins and minerals that are needed in very small quantities for different body functions. Examples include salt and the vitamins and minerals found in leafy vegetables, fruit, and vitamin supplements.

Carbohydrates

Carbohydrates are the body's prime energy source. They are found in bread, pasta, rice, potatoes, fruit, and candy.

Fats

Fats are essential for processing some vitamins, promoting healthy cell function, and are a rich source of energy.

Protein

Protein builds up, maintains, and replaces body tissues. It is essential for muscle growth and a healthy immune system.

Healthy eating

This is a graphic guide to a healthy meal. You should have more fruit and vegetables than any other food group. Cut down on big portion sizes, extra fats, and foods that are high in sugar and salt.

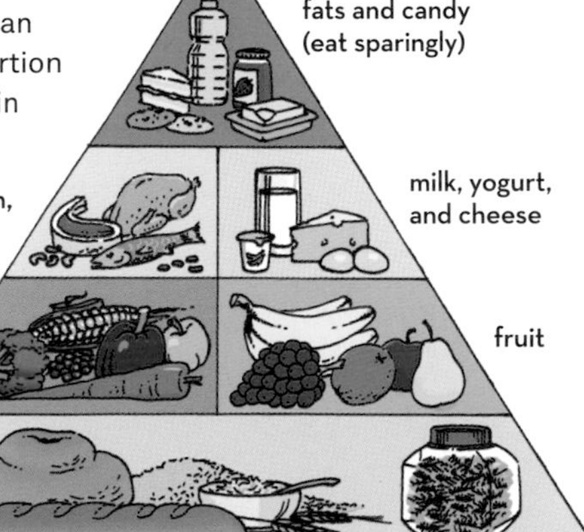

FOOD FOR THE OUTDOORS

A healthy diet is essential when you are enjoying the outdoors. When planning a trip, choose food that is healthy, tasty, lightweight, and doesn't need to be kept cold.

Hiking fuel

If you're on a big hike, your body will need a lot more food than normal. Graze on these easily digested, energy-rich foods to keep you going for hours.

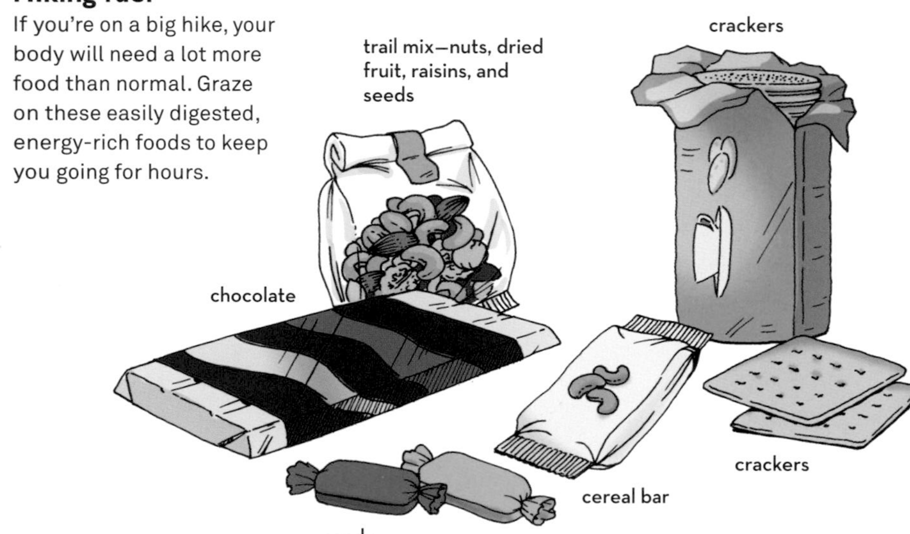

Breakfast

Lunch

Dinner

Dessert

FIRE MAKING

Humans have been making and cooking on campfires for a few hundred thousand years. Making fire is still an important skill to learn so that you can keep warm and cook when camping.

Fire triangle

There are three elements that must be present for a fire to exist: oxygen, fuel, and heat. You'll need them in the right combination to get your fire started. Removing one or more of these elements will put out the fire.

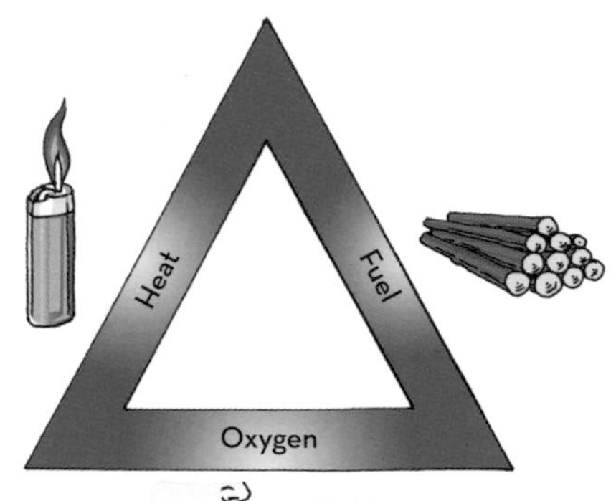

Tinder

Tinder is a fine flammable material that easily catches a spark.

Bark
Look for dry inner bark from dead logs.

Moss
Dead, dry moss makes an excellent fire starter.

Grass
Break down stalks of dry grass into fine fibers.

Fungus
The inner flesh from bracket fungus is flammable.

Cotton ball and petroleum jelly
A highly flammable mix.

Leaves
Dry dead leaves are often easy to find.

Parts of a fire

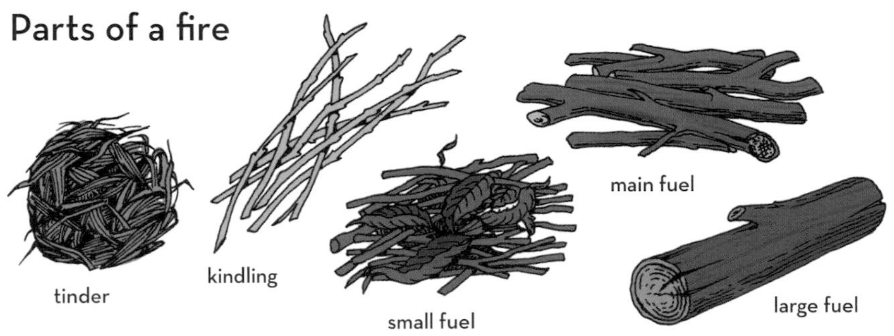

Build it up

A good fire is built up gradually. Start with tinder, then once the tinder has begun to burn, add kindling—dry twigs and sticks no thicker than your little finger. As coals are created, slowly add larger pieces of fuel.

Starting structures

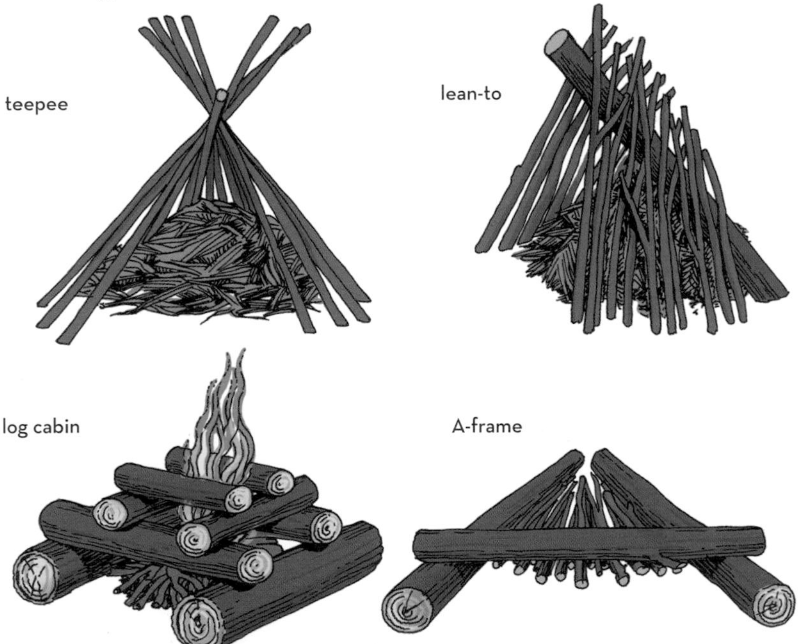

teepee

lean-to

log cabin

A-frame

FIRE STARTING

Starting fires has been a straightforward task ever since the invention of matches and lighters. However, there are other ways to create a spark if you don't have these tools available.

BEAR SAYS

Flint can be used to start fires even when wet, making it an incredibly useful piece of gear.

Heat sources

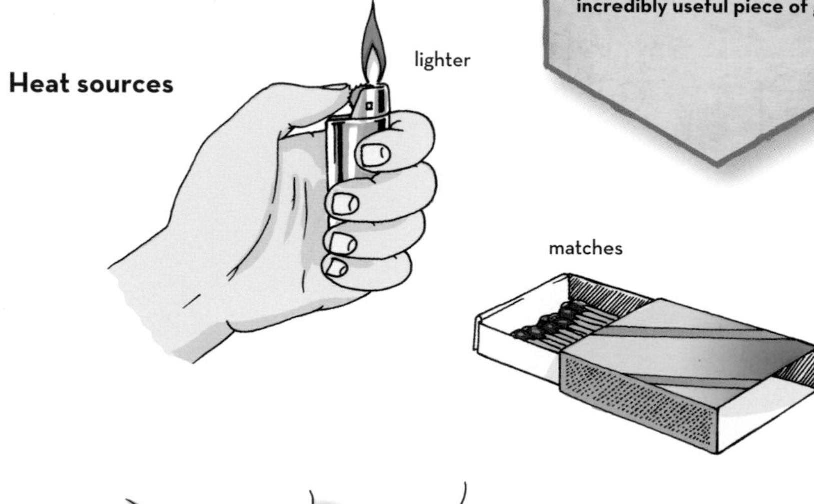

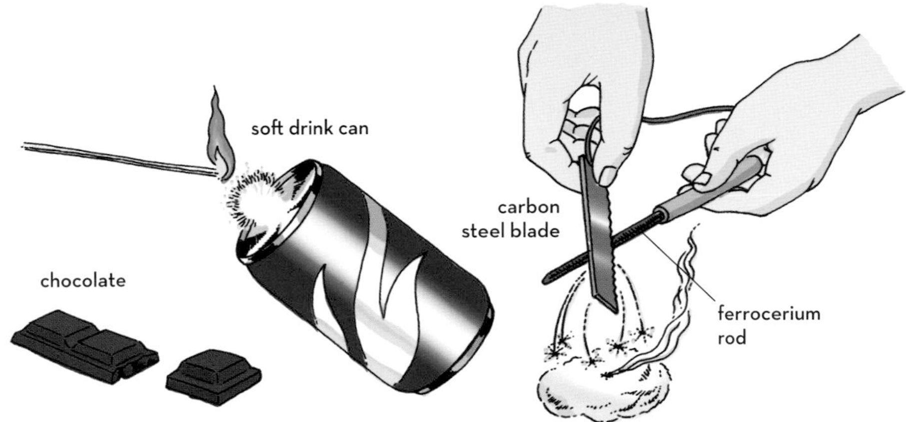

Parabolic can

Polish the base of a can with chocolate or toothpaste until it is mirror smooth and highly reflective (this may take several hours).

Flint and steel

The "flint" component of a flint and steel fire-starting kit is actually made of a metal alloy called ferrocerium. When struck with steel, it gives out sparks.

Battery method

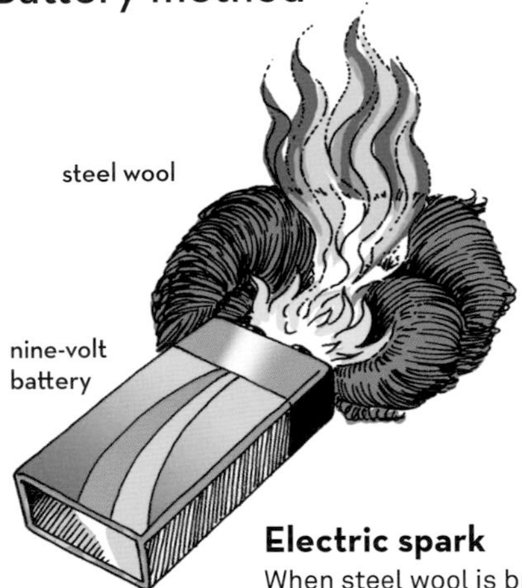

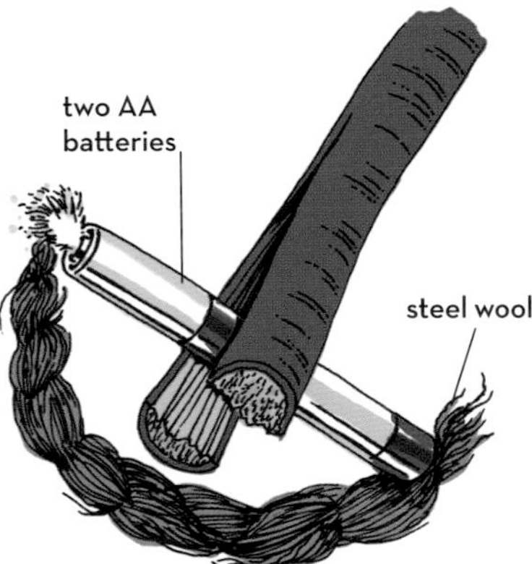

Electric spark

When steel wool is brushed against the contacts of a battery, it will glow brightly and begin to burn. A nine-volt battery is most convenient for this method, but any battery will work, including one from a cell phone.

Magnesium fire block

These fire-starting kits consist of a steel striker and a block of magnesium with a ferrocerium rod fixed down one side. They are small and light, and are still effective in damp conditions.

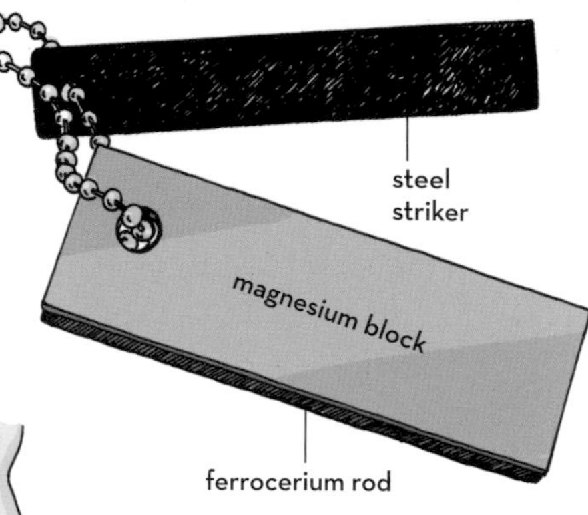

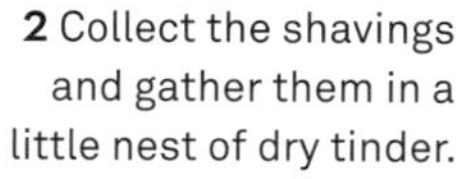

1 To begin, use a knife blade to scrape a small pile of shavings from the magnesium block. The shavings are light, so protect the pile from the breeze.

2 Collect the shavings and gather them in a little nest of dry tinder.

3 Run the ferrocerium rod along the steel striker or a knife blade. The resulting sparks will catch in the magnesium shavings and burn a very intense, white-hot flame for a few seconds—long enough to get your tinder burning.

Hand drill

A hand drill consists of a softwood drill and fireboard. Run your hands down the drill as you spin it, to maintain pressure and build friction.

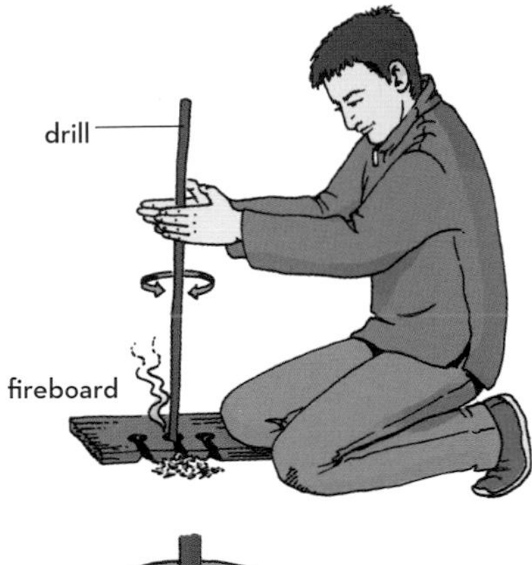

Fire piston

This ancient device is from Southeast Asia and the Pacific. Quickly pushing the piston into the cylinder causes a spark to be ignited in the tinder.

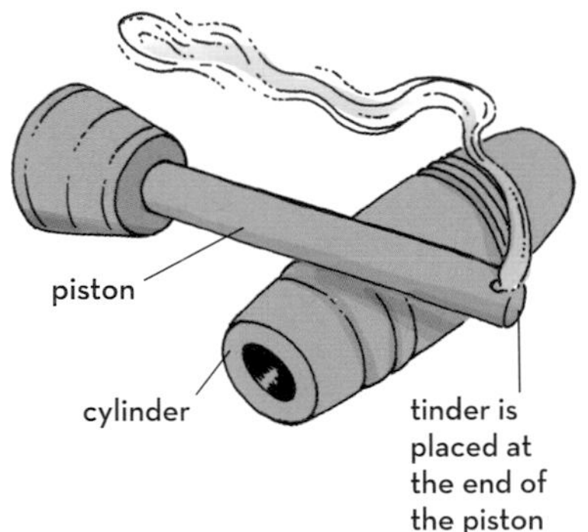

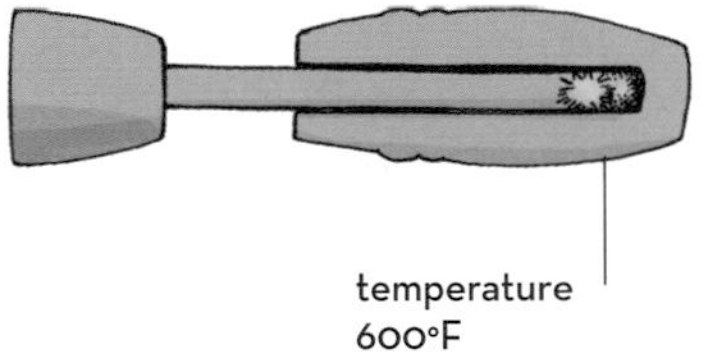

Fire plow

Cut a straight groove along a softwood base. Plow the tip of a hardwood rod back and forth along this groove. As friction builds up, small wood fibers will become detached from the groove. Eventually the detached fibers will start smoldering and form a "coal." Use this to ignite your tinder.

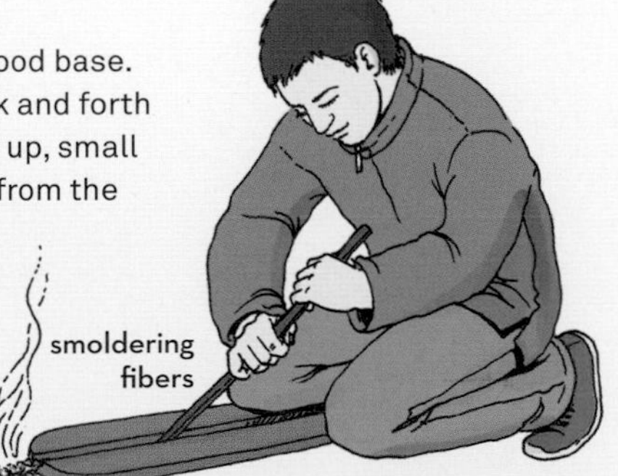

Bow drill

The bow drill is slightly more complicated than other friction fire-starting methods, but once mastered, it is extremely effective—even when temperatures are cool or weather conditions are humid.

BEAR SAYS

This device has been used since prehistoric times. It is still a very effective way of creating fire and learning how to use it could be invaluable.

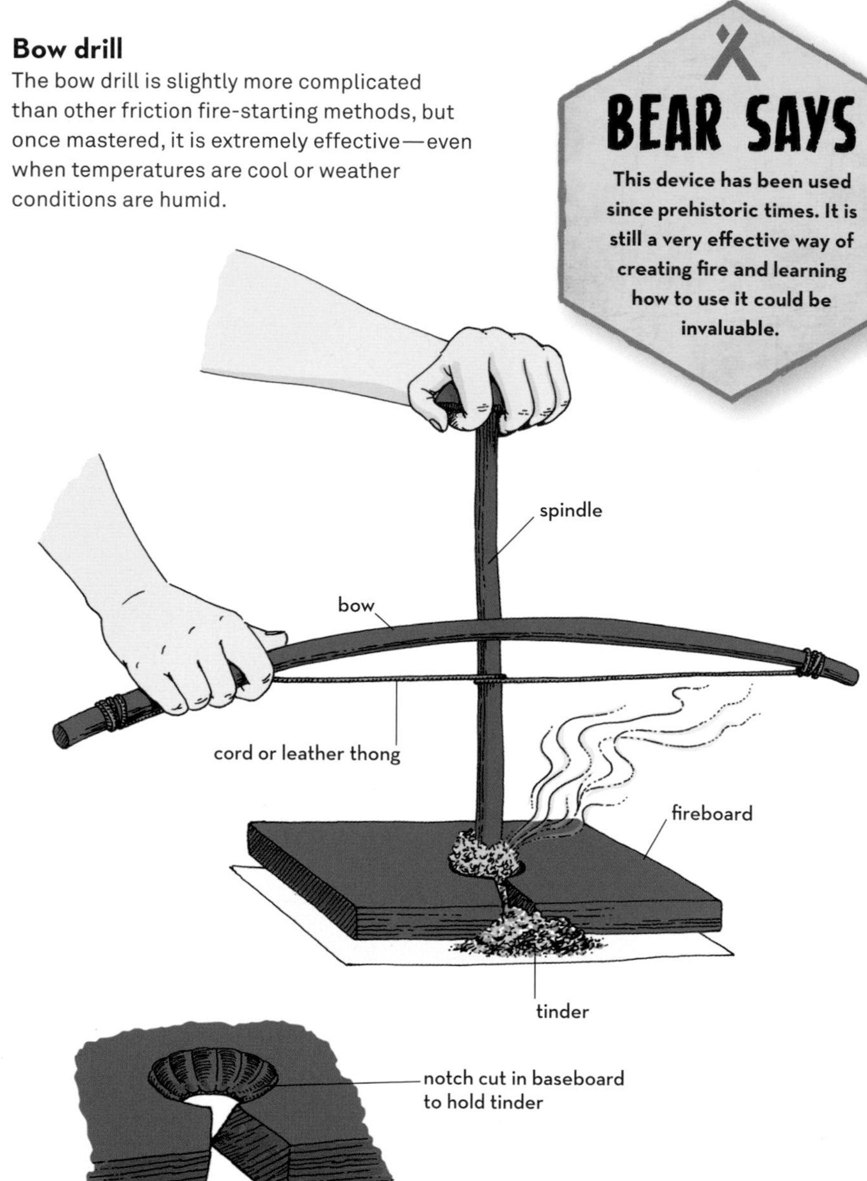

Blowing tinder

The end result of many fire-starting methods is not a flame but a precious glowing ember. To really get the fire started, quickly gather the ember into a bundle of tinder and blow gently. This adds oxygen and raises the temperature enough for the material to burn.

Carrying fire

It can be easier to carry embers rather than to start a fire without matches or a lighter. To do this, punch a few holes in a can and attach a cord or wire for a handle. Then place the embers between two layers of dry moss. Check the embers from time to time, and blow on them if they are starting to fade. Well cared for, the embers should last several days.

COOKING WITH FIRE

As well as providing warmth and a place to gather, the main purpose of a campfire is to cook your food.

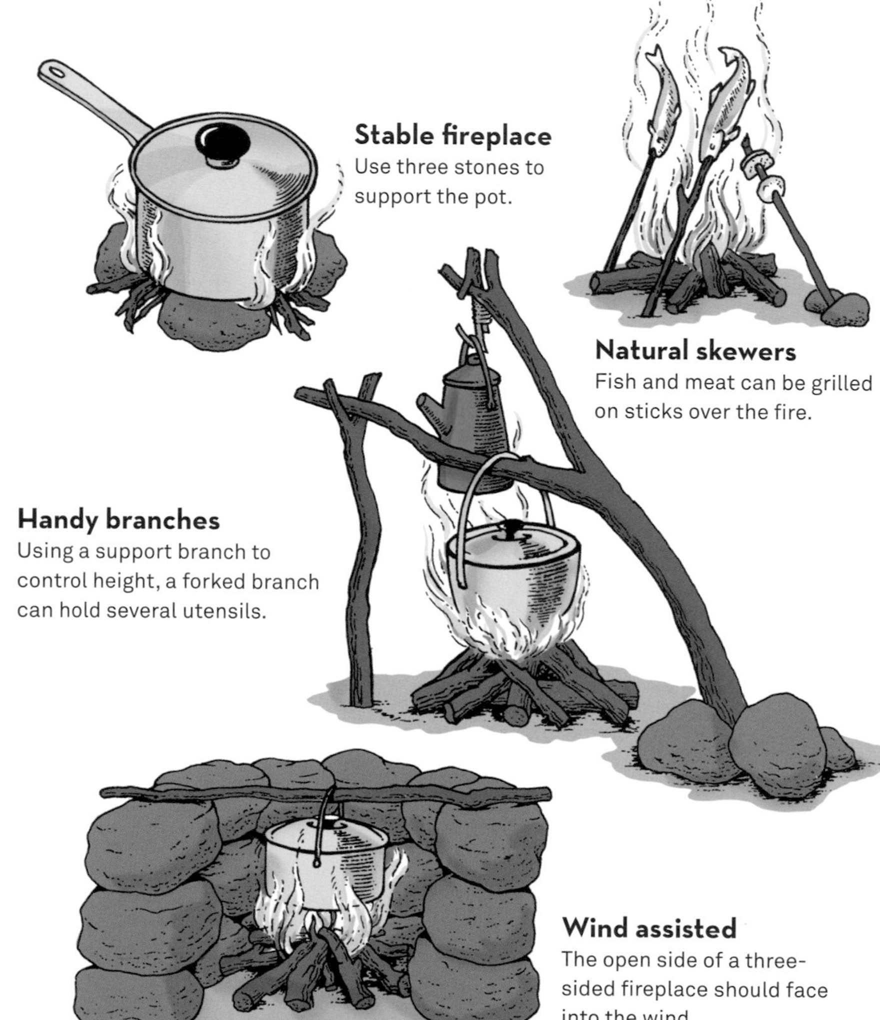

Stable fireplace
Use three stones to support the pot.

Natural skewers
Fish and meat can be grilled on sticks over the fire.

Handy branches
Using a support branch to control height, a forked branch can hold several utensils.

Wind assisted
The open side of a three-sided fireplace should face into the wind.

Basic construction

Put two logs parallel to the wind to form a simple fireplace.

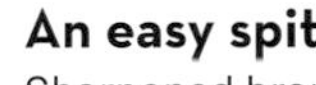

An easy spit

Sharpened branches driven into the ground offer a sturdy spit.

Uneven surface

Use the slope of the ground and some large rocks to help support your utensils over the fire.

Longer term

If you are staying in one place for a while, dig a hole for a more permanent fireplace.

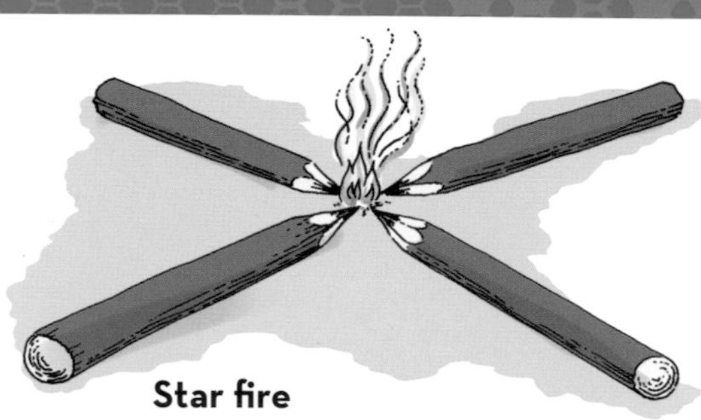

BEAR SAYS

As well as for cooking, fire is needed to heat water. Water should be boiled to make sure it is safe to drink, and it's preferable to wash in warm water.

Star fire

Push in the logs as they burn to create a long-lasting cooking spot.

Crane

This arrangement will keep your cooking pot off the fire, and keep it from getting smothered.

Adjustable crane

This crane allows you to move your pot up and down so that you can control the temperature it is exposed to.

Stone griddle

A slab of stone will take a long time to heat, but will stay hot for a long time. Use a dry, solid rock.

Bamboo cooking pot

Green bamboo is very fire resistant and makes an excellent pot for boiling and simmering.

food goes here

steam

water

Bamboo steamer

Punch a few holes in each of the two walls that divide a length of bamboo into three sections. Put water in one end and food in the other and you have a steamer.

Breakfast in a bag

Line the bottom of a damp paper bag with bacon, then crack an egg on top. Place the bag on some hot coals and ashes to cook.

Foil oven

Wrap a whole meal in tinfoil and put it into the coals for a slow roast. By using this method exclusively, you can save on the weight of cooking pots and pans.

Hangi

The hangi is a traditional New Zealand Maori method of cooking large communal meals. To make a hangi, first dig a pit in the ground. Then build a pyre of wood beams over the pit to carry the hangi stones. Set the pyre ablaze to super-heat the stones. Once they have dropped into the pit, add the food in wire baskets, cover in damp sacks and soil, and leave to cook for two to three hours.

Mud baking

1 Gut a fish and lay it on a bed of non-poisonous green leaves. There is no need to remove the scales.

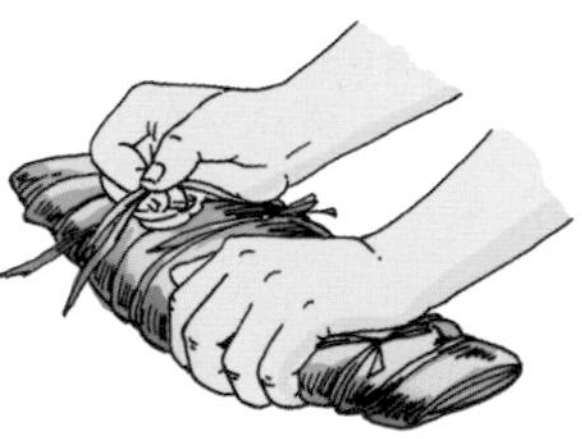

2 Fold the leaves over the fish, ensuring that it is completely covered. Bind the package with twine.

3 Pack mud all around the package. Use clay if it is available, or use mud that has a claylike texture. Check for holes.

4 Bury the package in hot coals. A medium-sized fish should take about 20 minutes to cook.

CAMPING STOVES

Cooking meals on an open fire has its downsides. Things get sooty, and it can be hard to find fuel. Often, a camping stove is the best way to cook outdoors.

Fuel stoves

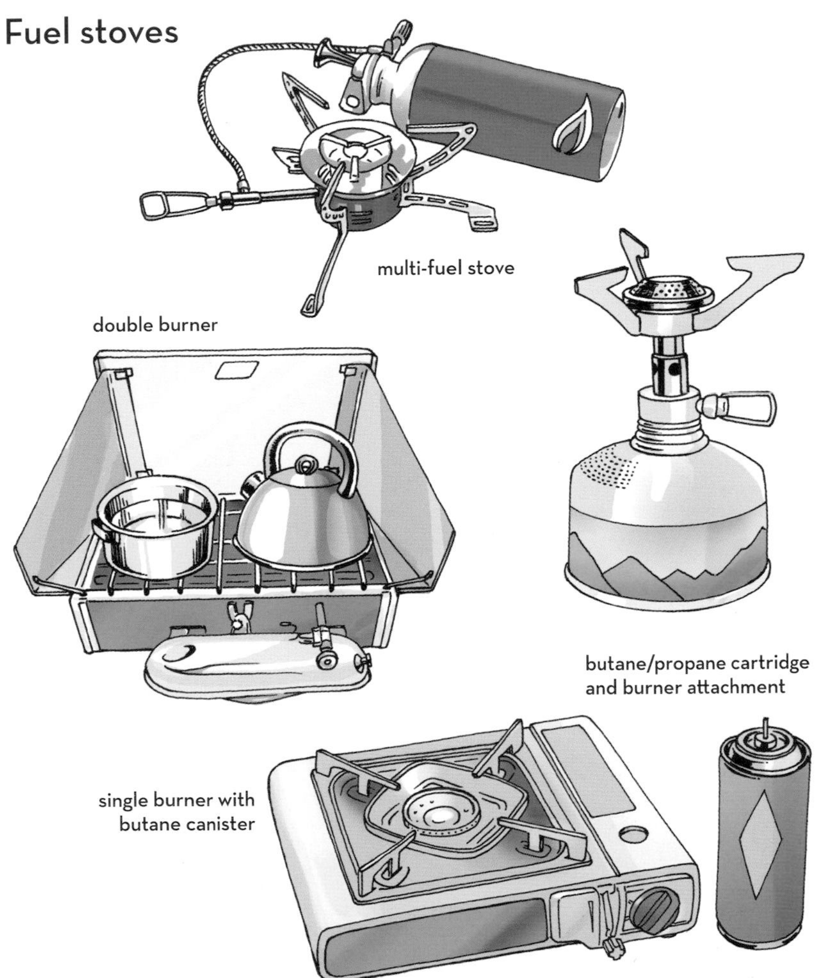

Alcohol burner set

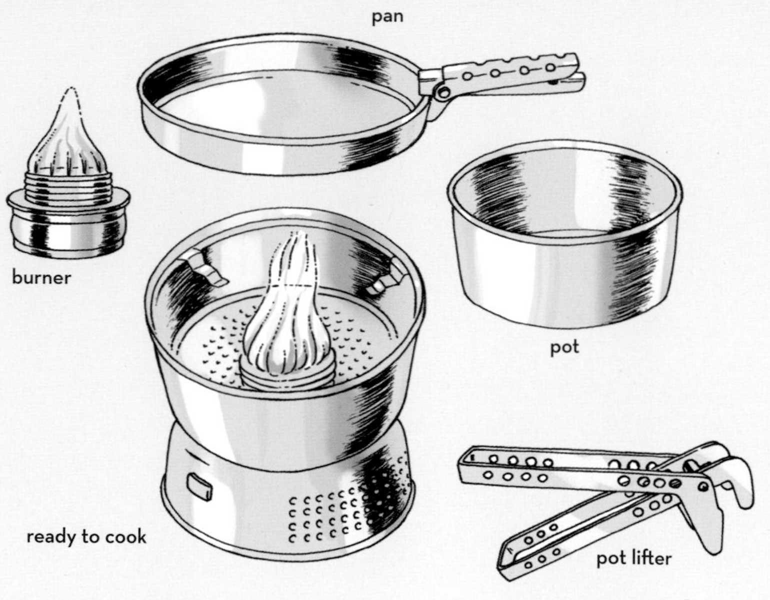

Primus paraffin stove

Invented in 1892, the Primus pressurized paraffin burner was the original camping and expedition stove.

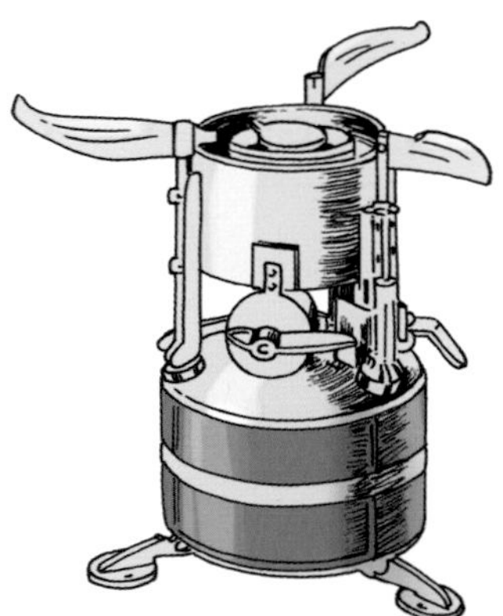

US Army gasoline-burning stove

These stoves were standard US Army issue from 1951 until 1987.

Cooking without gas

Parabolic solar cooker

Solar cookers are the ultimate in environmentally friendly cooking. Parabolic cookers can reach high temperatures very quickly and are good for bringing liquids to a boil.

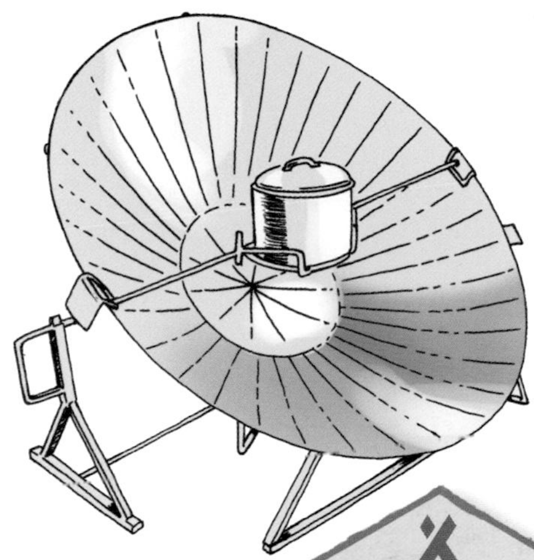

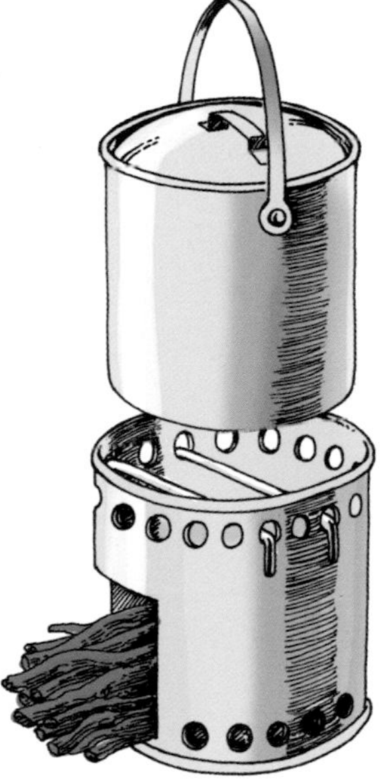

Twig stove

This low-impact twig burner can be easily made from a steel can and wire using basic tools.

BEAR SAYS

If you are prepared, cooking in the wild can be easy. Be extra careful with camping stoves, however, as they can reach extreme temperatures.

Solar box oven

This design makes use of reflective panels and a sealed light-absorbent chamber. Temperatures inside can reach 300°F.

MAPS AND NAVIGATION

Learning essential skills such as map reading and navigation will allow you to truly experience the world around you. Although it requires practice and practical application, there is nothing like the sense of achievement you will feel upon successfully finding your way to your planned destination!

IN THIS SECTION:

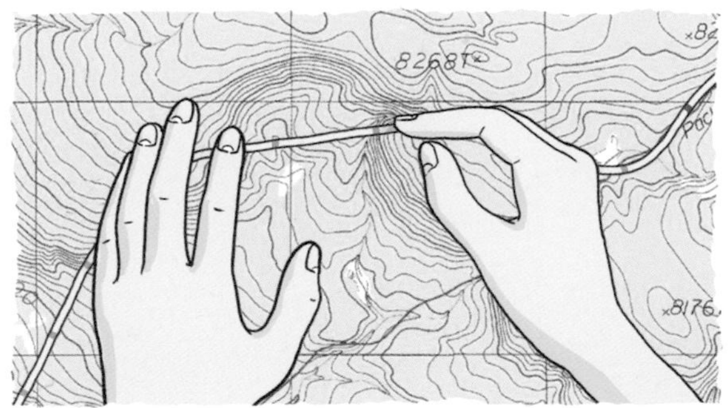

STARTING YOUR ADVENTURE

The world is full of exciting places to explore. In order to stay safe on your travels you will need to learn the art of navigation and map reading, like the adventurers of the past. As long as you have a map and compass, and know how to use them, you need never get lost. Remember, practice makes perfect!

Equipment

If you venture into the wild, make sure you have all the equipment you need so that you can get from one place to another as quickly and easily as possible. The type of equipment and the amount you need depends upon where you are going and for how long. Here are some of the basics you will need when going walking.

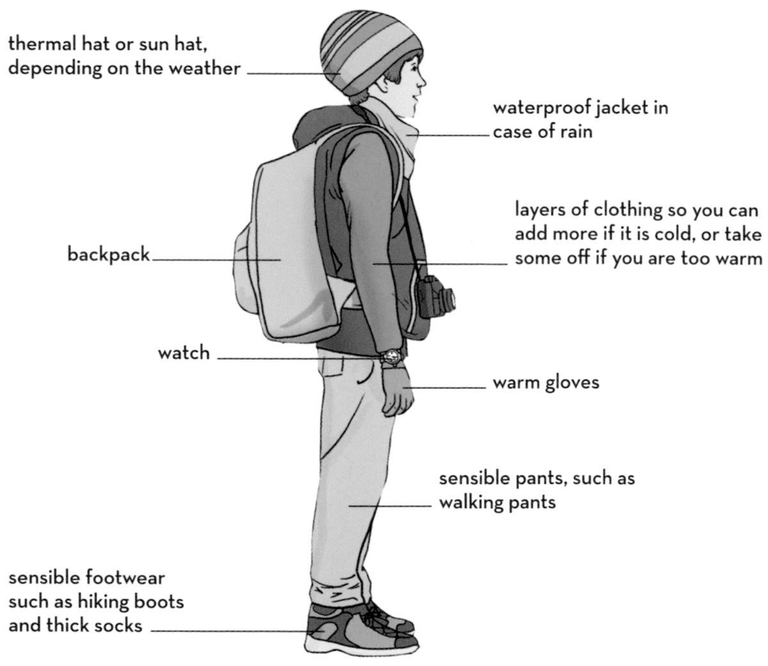

Make sure you carry these essentials in your backpack.

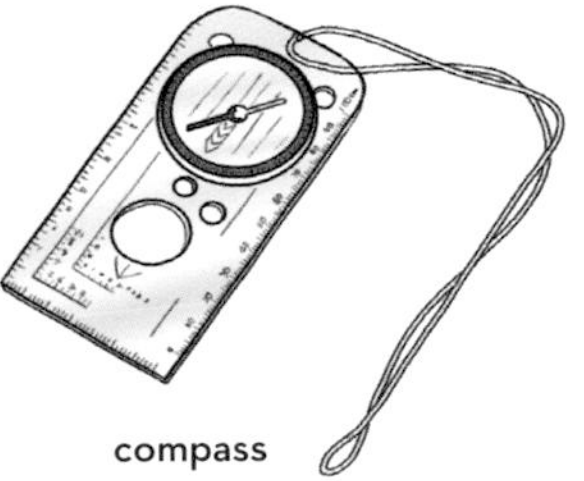

compass

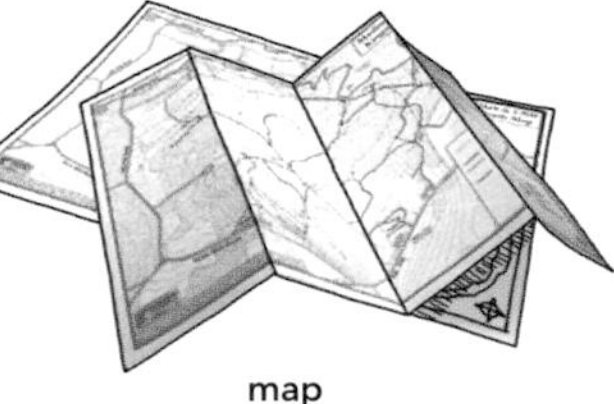

map

first-aid kit

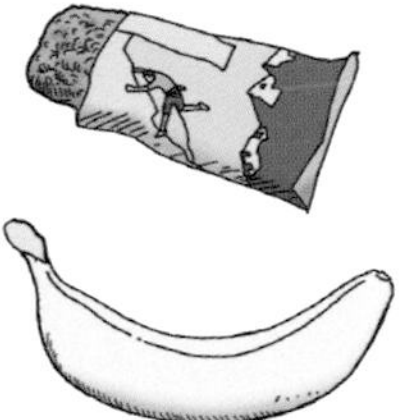

food and bottle of water

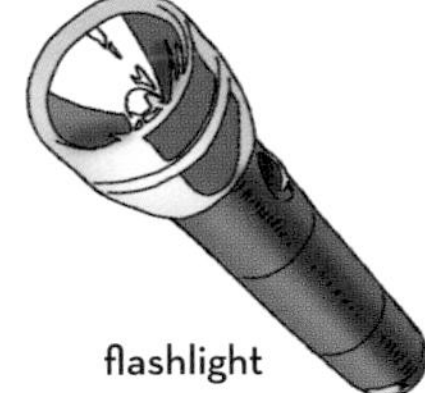

flashlight

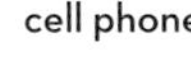

cell phone

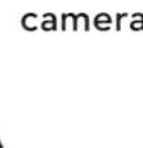

camera

sunscreen and insect repellent

emergency contact details

BEAR SAYS

Make sure someone knows your route and estimated arrival time before you set off so they can raise the alarm if you don't turn up when expected.

MAPS

Map care

As you walk, you will need to look at your map quite often to check that you are on the right track. Maps need to be folded carefully to keep them in good condition so that they stay useful.

A map case with a neck cord may be a good idea, while some outdoor clothing has a specially designed large map pocket—if you are keeping the map in a bag, make sure it is easily accessible when you are out. Laminated maps can be useful if you are likely to be outside in wet or windy weather conditions. A spare may also be essential if you are somewhere remote.

How to fold a map

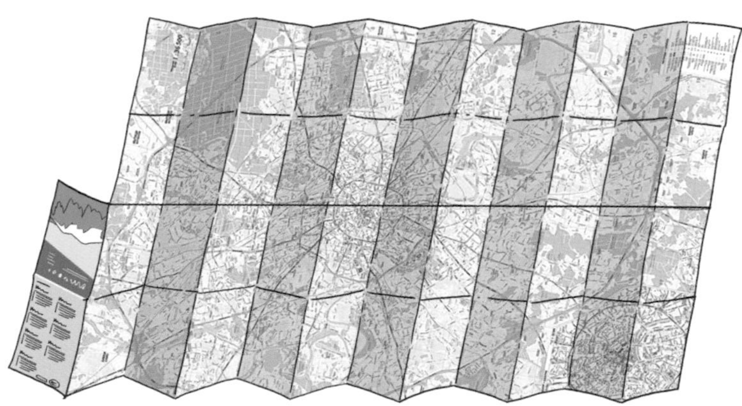

1 Although it sounds easy, folding a map can be tricky. With your map spread open, look at the creases—they should show the correct places to fold.

2 Fold the map in half by bringing the top edge to meet the bottom edge.

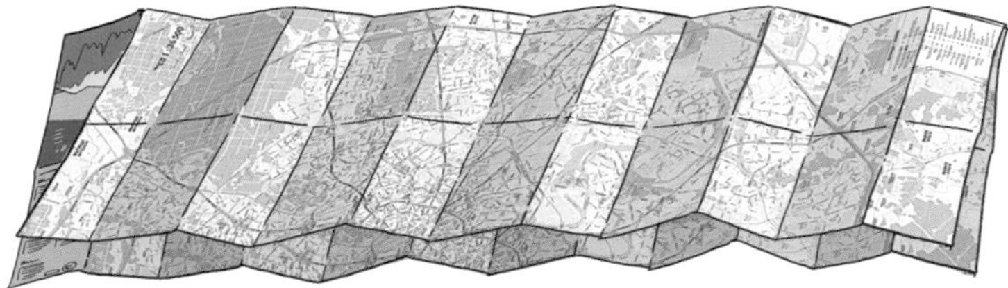

make sure the cover is out to one side

3 Fold the map inward, as though it is an accordion.

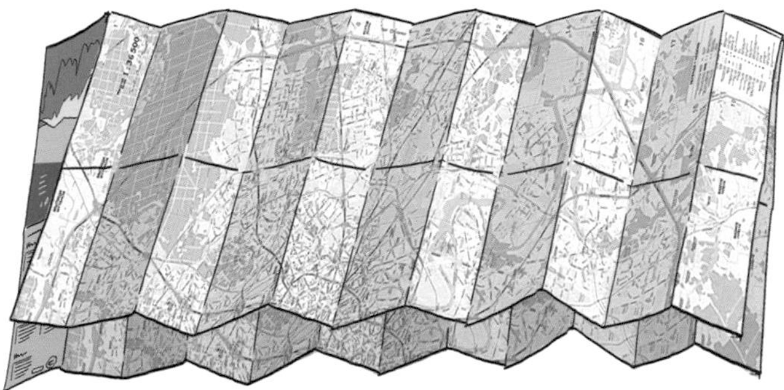

4 Fold the map over so that the cover is on top. Some maps have a third section that needs to be tucked in.

BEAR SAYS

Explorers should be good at drawing maps, so it is important to get lots of practice. You never know when you might need these skills!

Topographical maps

These maps show contour lines and landforms. They are useful because they show how the ground is shaped. Every hiker should be able to "read" the contour lines on the map so that they can plan their route.

Contour lines

Each contour line on a map joins up points where the ground is the same height. Lines close together mean that there is a steep slope, while lines that are far apart indicate flat ground or a gentle slope.

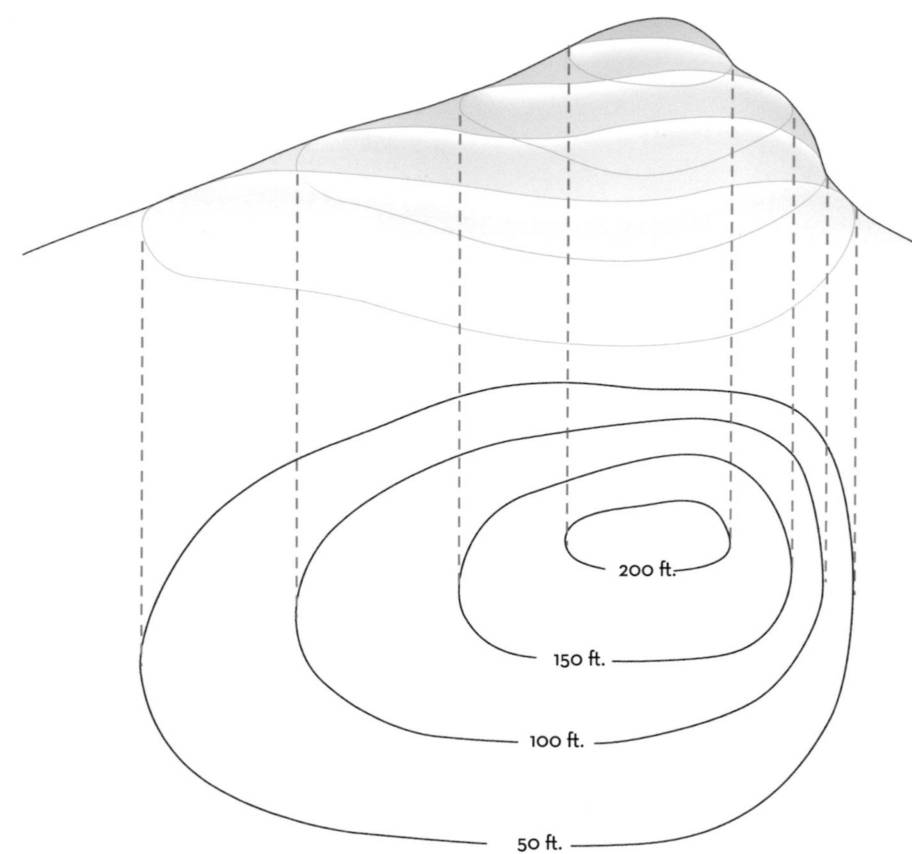

Choosing the best route

The best route to a destination may not be the most direct. Choose a path that is safe—for example, on footpaths—and stay away from any hazards like cliffs and fast-moving water. Make it as simple as possible so that it is easy to follow and make sure you have permission to be on the land if it isn't a public right of way.

Landmark spotting

From your known start point, look for the nearest and biggest landmark on your map, such as a lake. Check the contour lines in case it is hidden from view. If you don't know where you are, mark any notable landmarks around you on a piece of paper and then see if you can match them up to your map. The scale may be tricky to guess, but it is a good start if you are unsure of your location.

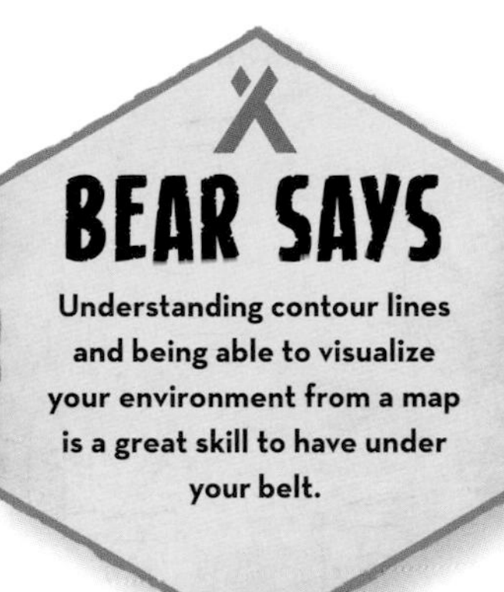

NORTH

WEST

EAST

SOUTH

How to measure distance

Map measurers allow you to find the distances between two points on a map. Route planning is an important part of navigation, and knowing how far you need to travel will help you work out your finish time—and if you'll arrive before dark!

map scale

needle

Analog map reader

This tool uses a needle and dial printed with different map scales to show distance in miles or kilometers.

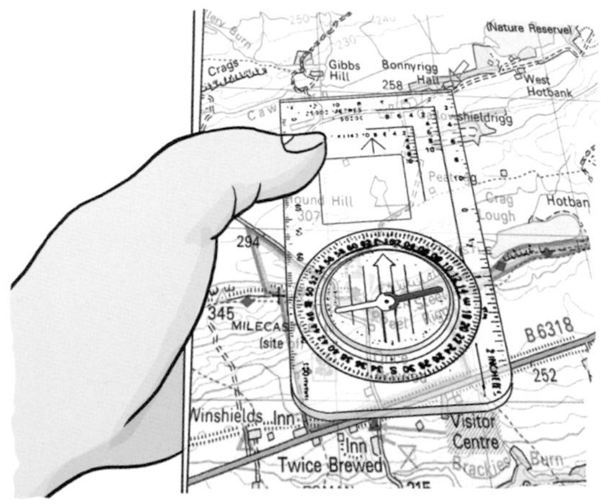

Using your compass

As well as taking bearings and helping you find north, your orienteering compass can help you find distances, using the measurements on the base plate.

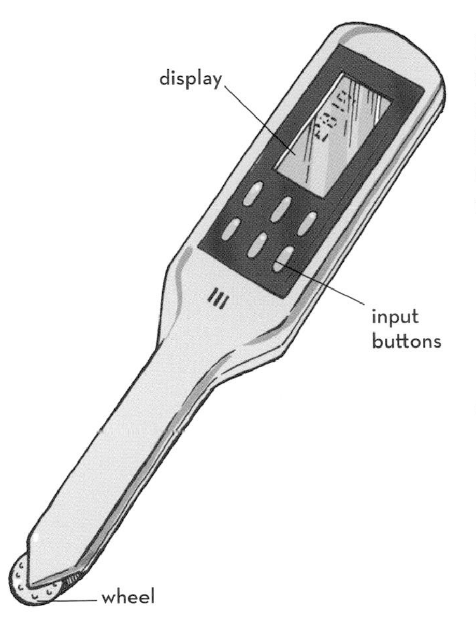

Digital map reader

With digital map reader wheels, the user sets the unit, such as kilometers, and the map scale. They then run the tip of the reader along the route to find the distance.

Measuring distance accurately will help you plan your hike. Take the time to take measurements before you set off.

Using a piece of twine

If you are without a device, you can place a piece of twine along your route, and then along the distance scale on the map.

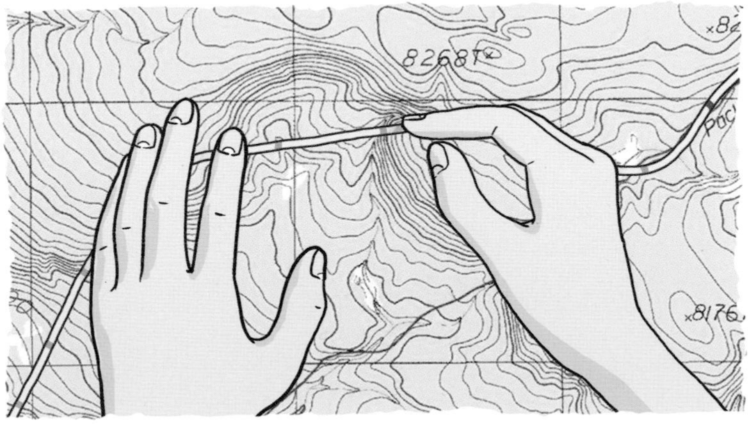

Types of landscape

As you walk around the countryside you will notice that there are many different types of landscape. Some are far easier to travel along than others. These are some of the most common types.

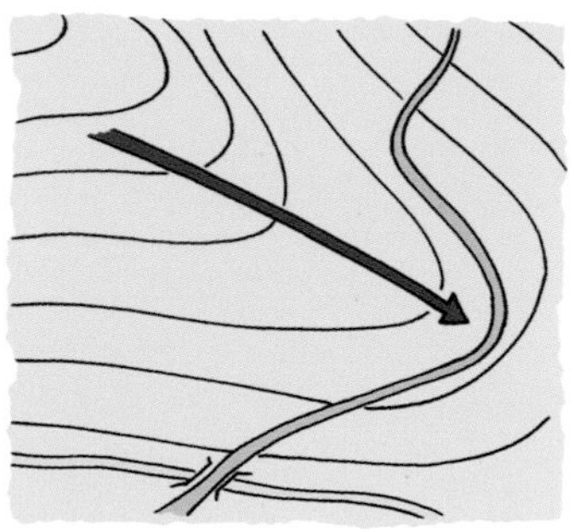

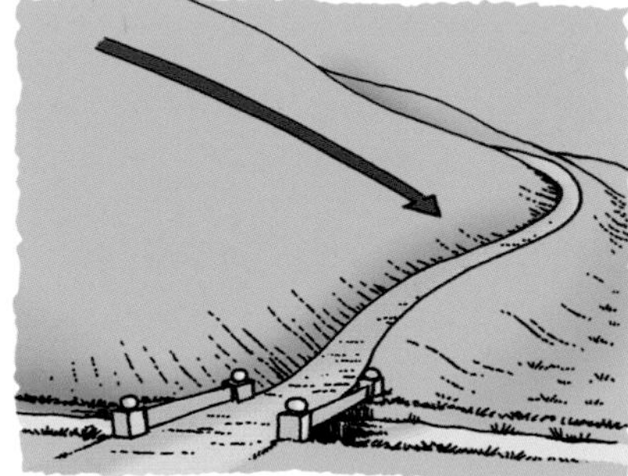

Gentle slope

This slope decreases steadily in height and makes for an easy climb.

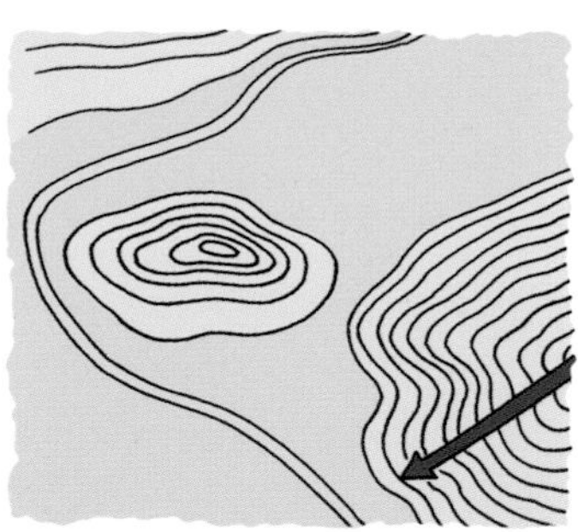

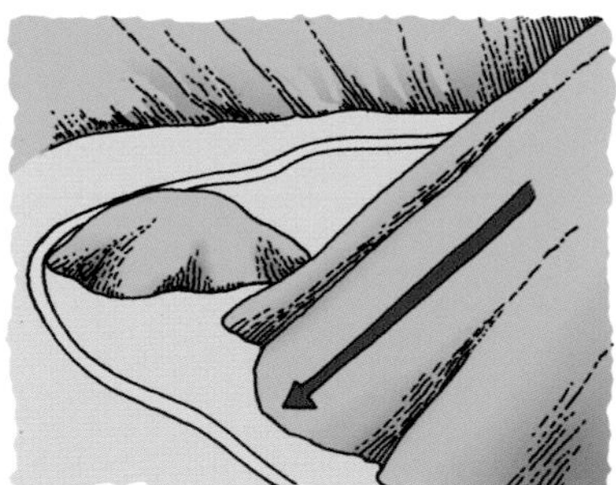

Steep slope

Steep slopes are hard to trek up, due to the sharp angle of the land.

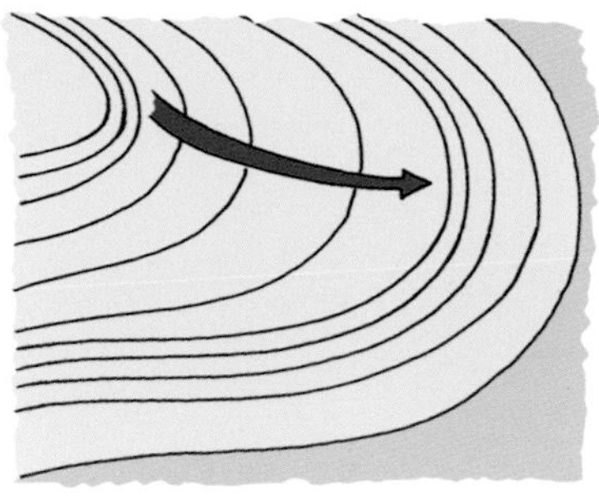

Concave slope

This kind of slope is steep at the top but less so at the bottom, a little like the curve inside a bowl.

Convex slope

A rounded slope that goes from less steep to steep.

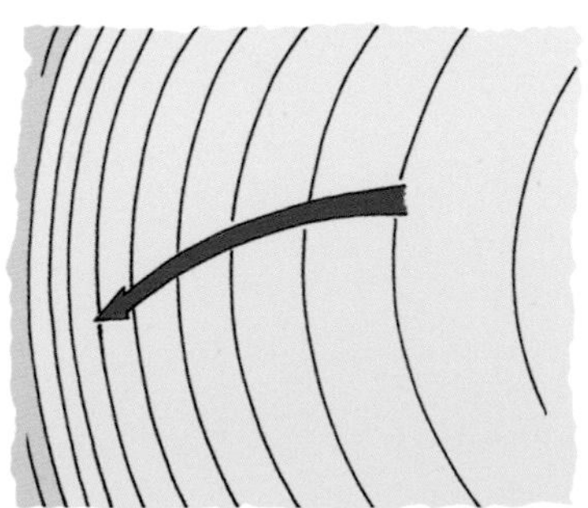

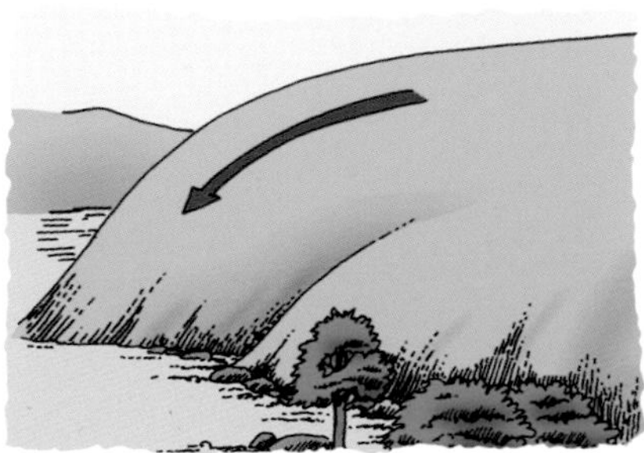

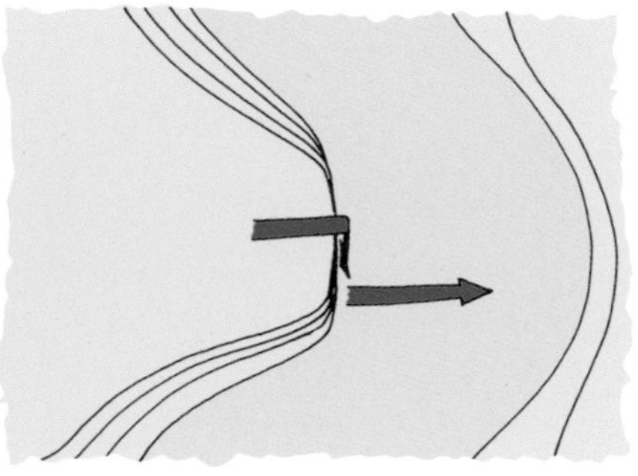

Cliff

A cliff is a very steep drop in the landscape, often at coasts.

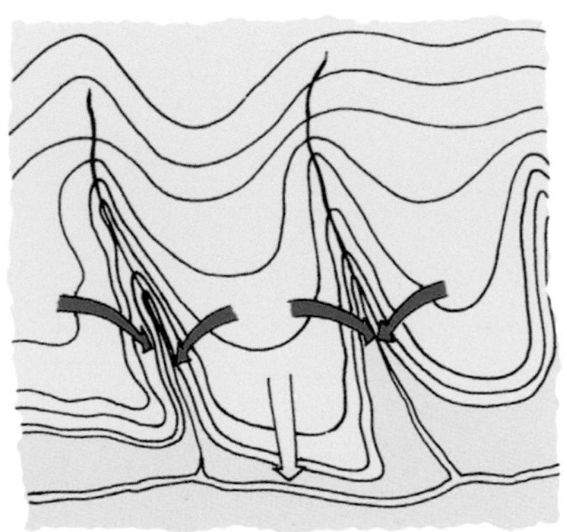

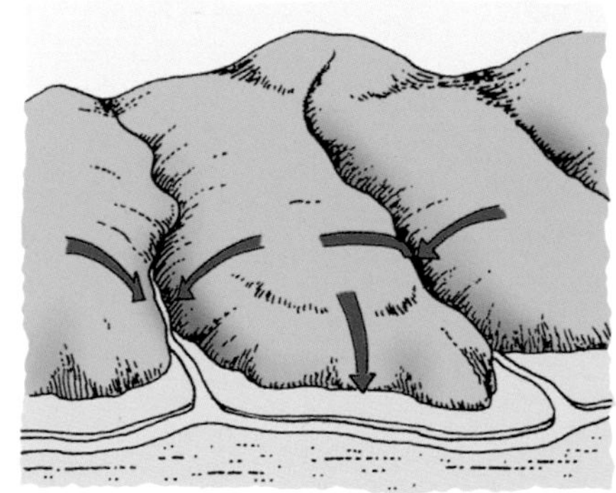

Gullies and spurs

Gullies are steep ravines found close to the seashore, while spurs are the ridges of land that slope down from the edge of a hill.

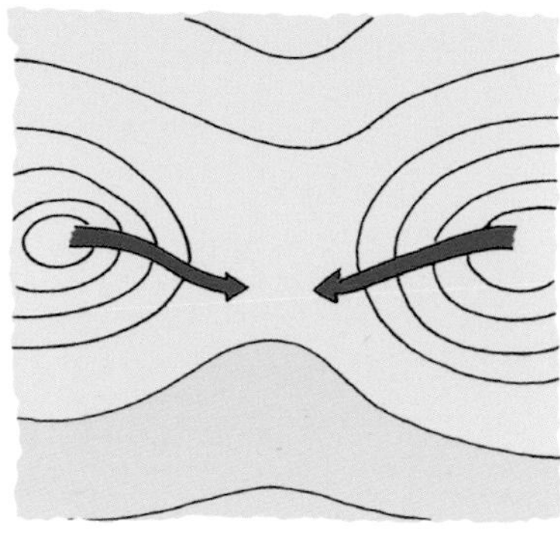

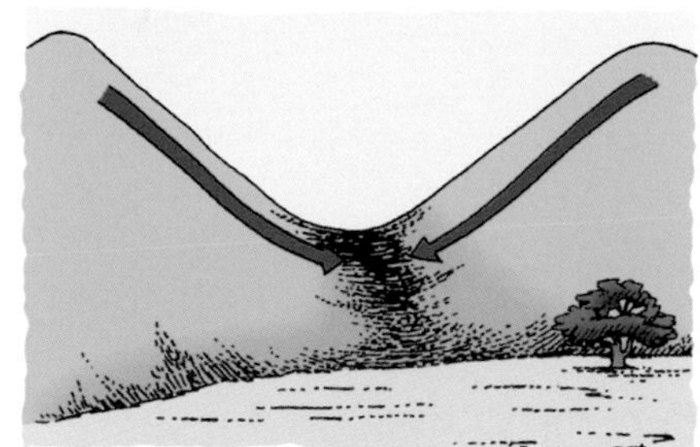

Saddle

The area between two connecting hills or mountain peaks is called a saddle.

Valley

A valley is a low area of land found between hills. Rivers are often found in valleys.

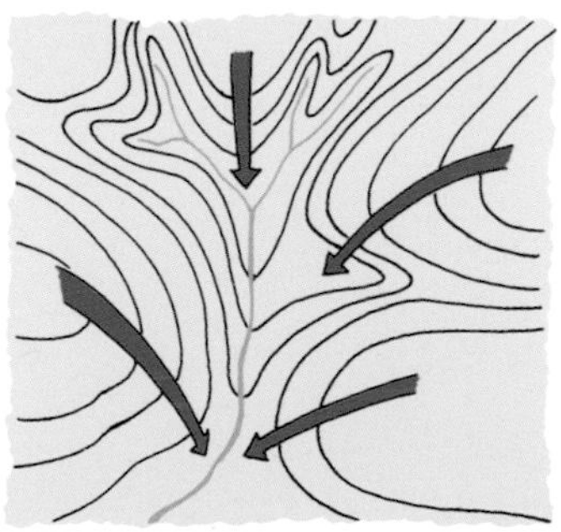

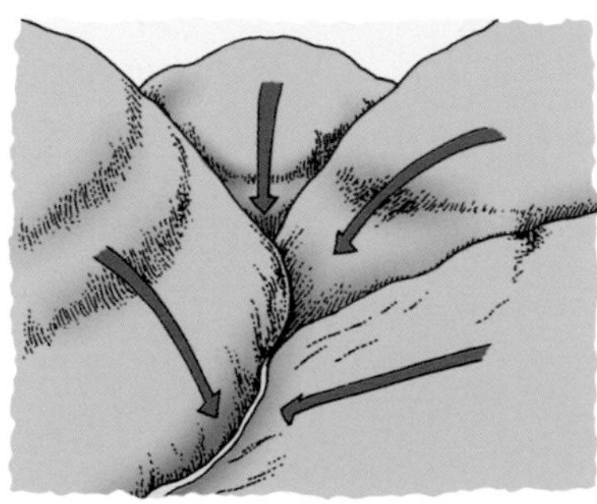

Symbols

Maps use symbols, lines, and colors to describe what is on the land and make the map clearer.

Learn the pictures

If you look closely at a map you can see it is covered in symbols. It would be almost impossible to write everything out in words on a map—there just isn't enough space. A key explains what these symbols mean. The symbols could be pictures, words, or abbreviations.

A symbol for Mars

Ordnance Survey is the national mapping agency for Great Britain and is one of the world's largest producers of maps. The symbols used on these maps are easily recognizable and in May 2016 they held a competition to design a symbol for Mars as it has been recently mapped.

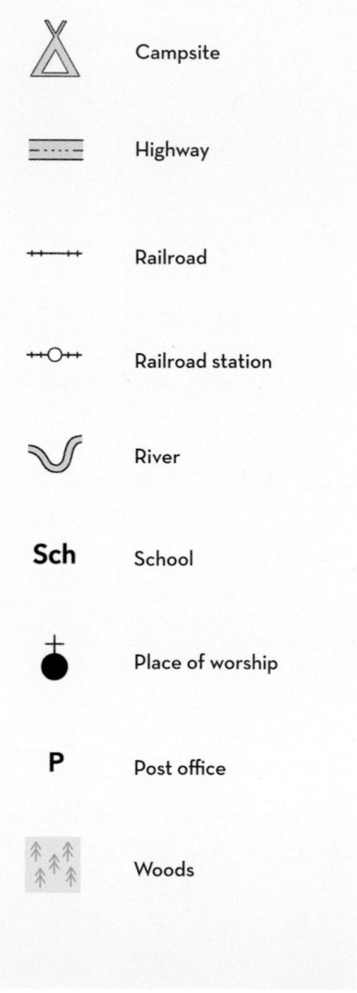

Scales

Maps are made to scale so that the distance between landmarks and places in real life is shown accurately on paper. It is very important that the map is correct, so that features are where the map reader expects. Scale also helps us to work out distances. Most maps have a scale written on them (e.g. 1:50,000). This means that one inch on the map represents 50,000 inches on the ground.

Small scale

The scale is shown visually on a map in both centimeters and inches. In this example, this diagram shows a map with a scale of 1:100,000 — every centimeter on the map is equal to a kilometer on the ground. The metric system of centimeters, meters and kilometers is more straightforward than the inches and feet we use in North America, so seasoned explorers usually work with the metric system. But you can work in inches and feet with a little bit of basic math!

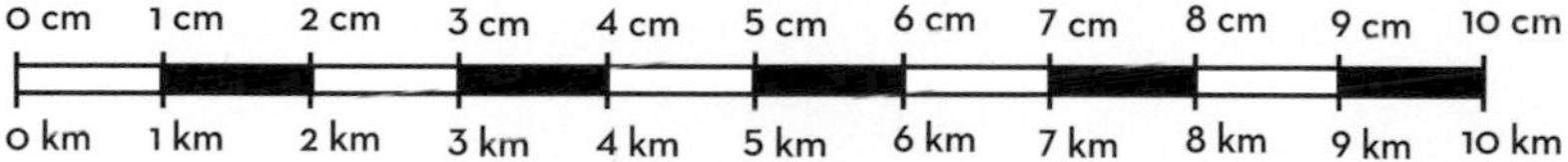

Coordinates

Coordinates can describe any location on Earth. Our planet is a globe or sphere, and around it are a set of imaginary rings drawn from east to west and north to south. The lines running from the top to the bottom of the globe are called lines of longitude, while the lines running around it are called lines of latitude.

Lines of latitude

Latitude is measured in degrees north or south of the equator (0–90°). The equator is an imaginary line that runs around the center of the planet, and is itself a line of latitude. It divides the Earth into two parts—the northern and southern hemispheres.

Lines of longitude

Lines of longitude run around the Earth from north to south. They are called meridians. Longitude is measured in degrees east or west of the Greenwich meridian (0–180°).

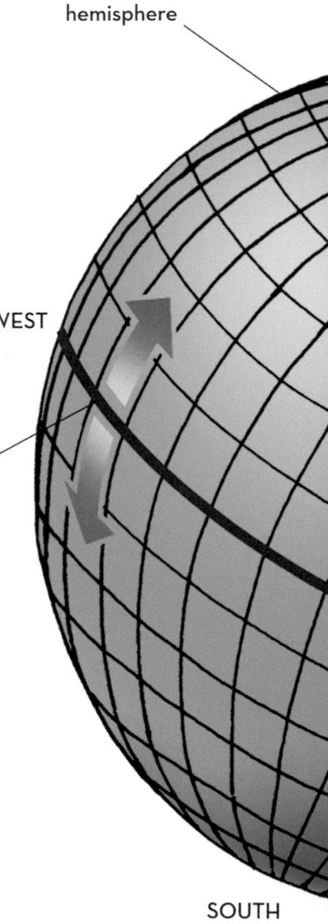

BEAR SAYS

Using longitude and latitude you can accurately describe any location on Earth.

Where do you live?

The coordinate system can be used to pinpoint anywhere on the planet, and works in any language because it relies on numbers. You can look up the latitude and longitude of where you live on the Internet.

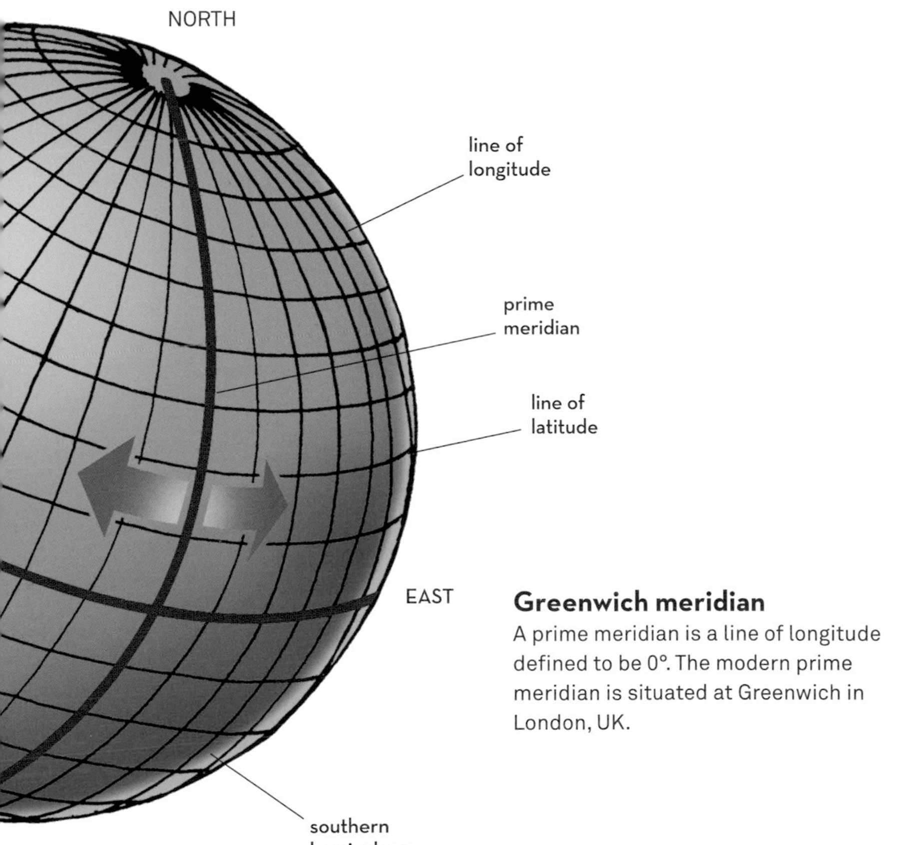

Greenwich meridian

A prime meridian is a line of longitude defined to be 0°. The modern prime meridian is situated at Greenwich in London, UK.

From globe to map

Our planet is a sphere—but maps are flat. In order to create a map, cartographers have to "project" the 3-D globe onto 2-D paper by squashing and stretching it. Luckily, the changes made to the globe are too small to cause big problems on most hiking maps.

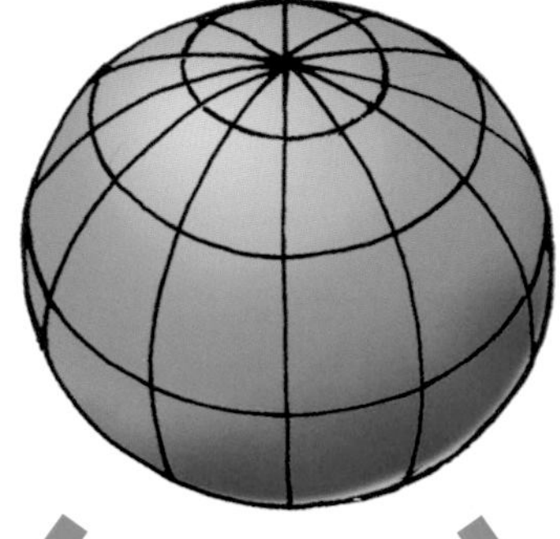

cylindrical projection

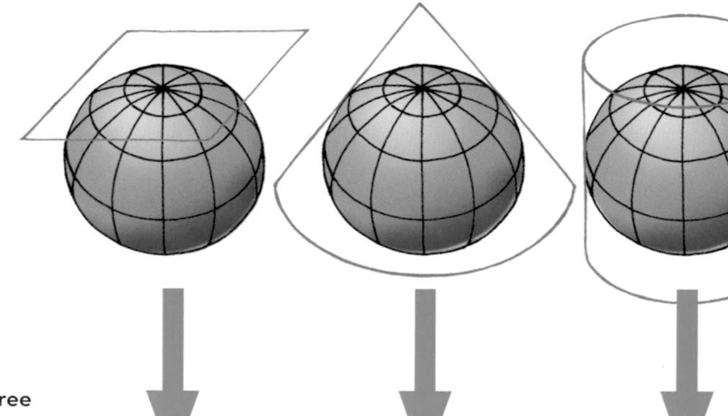

there are three main ways maps are projected onto flat paper

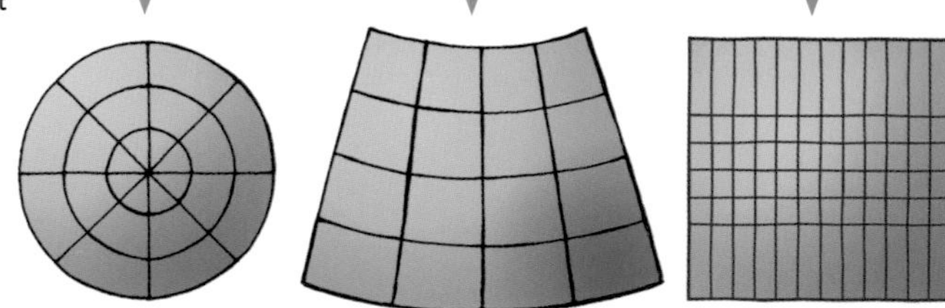

Grid references

Maps are often divided into squares called grids. These grids help to pinpoint a location on the map quickly. The vertical lines crossing the map from top to bottom are called "eastings" because the numbers go up as you move east across the map. The horizontal lines crossing the map from one side to the other are called "northings" as the numbers increase as you get farther north.

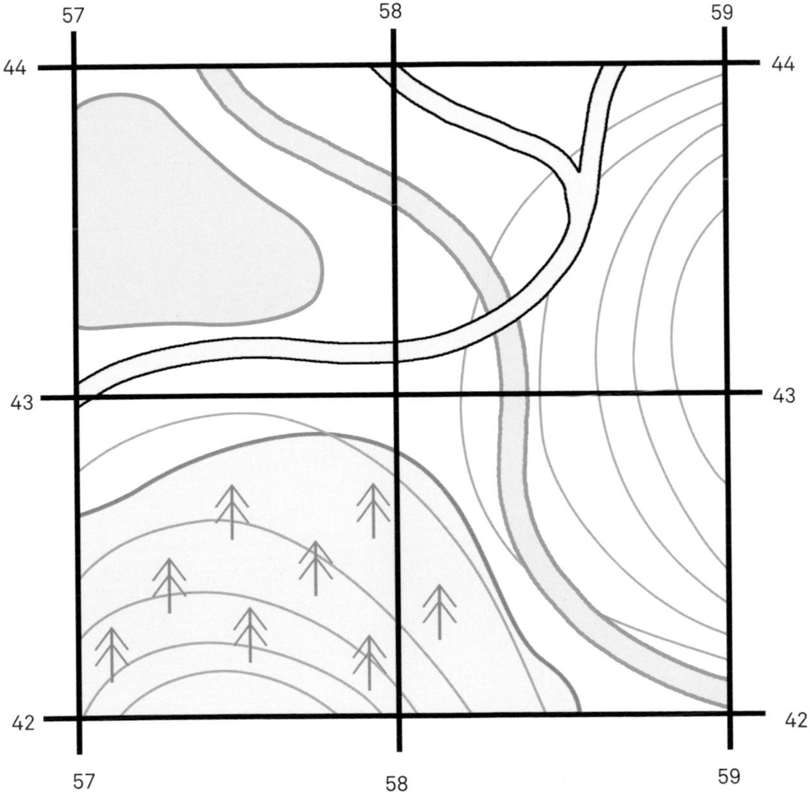

When you give a grid reference you always give the easting first. The forest is in 4257.

MAGNETIC EARTH

Our planet is one big magnet, which is very useful for navigation—just as long as you have a map, a compass, and an understanding of how magnetic north and true north differ.

Earth's magnetic field

At Earth's center is liquid iron. This metal core makes our planet into a giant magnet, with a magnetic field that changes over time. Because magnetic north is not at the geographic North Pole, you need to make special adjustments when using a compass. This is known as declination.

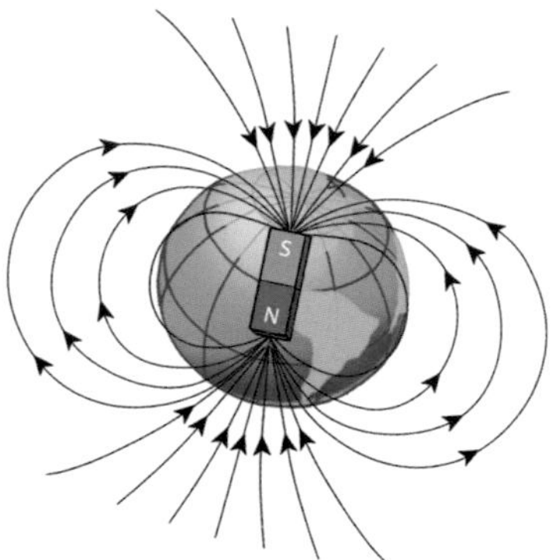

Earth's magnetic field is strongest near the North and South Poles.

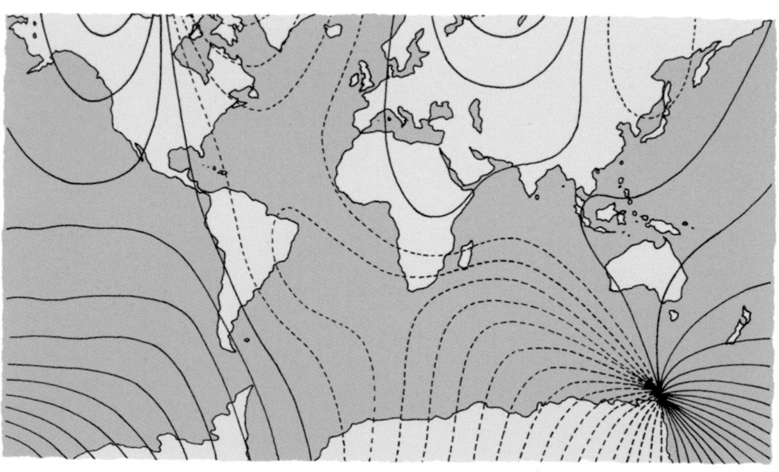

Which way is north?

When we talk about "true north" we mean the top of Earth—the geographic North Pole. Earth spins around, and if you imagine it is spinning around a line that goes from the top to the bottom of the planet, the North Pole is the point at the very top, and the South Pole is the point at the very bottom. Magnetic north is different as it responds to Earth's changing magnetic field, and doesn't line up with the same point.

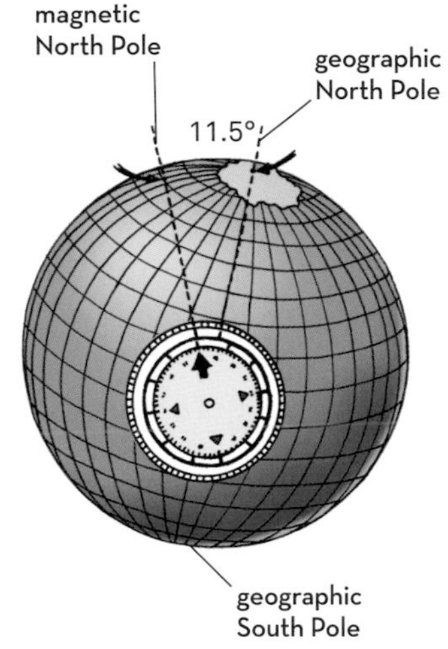

Three norths

Good maps will have a key to indicate grid, true (geographic), and magnetic north.

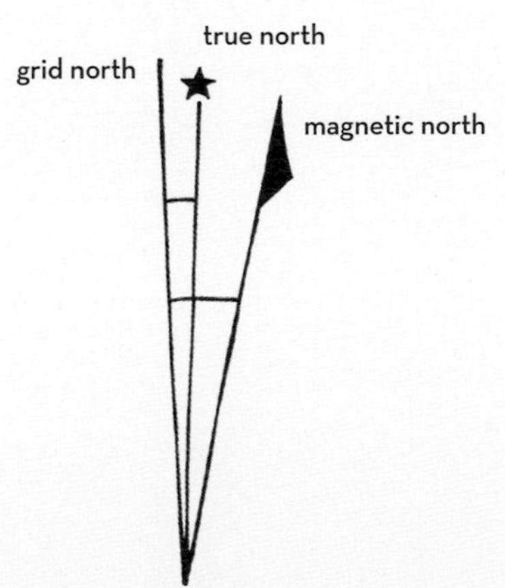

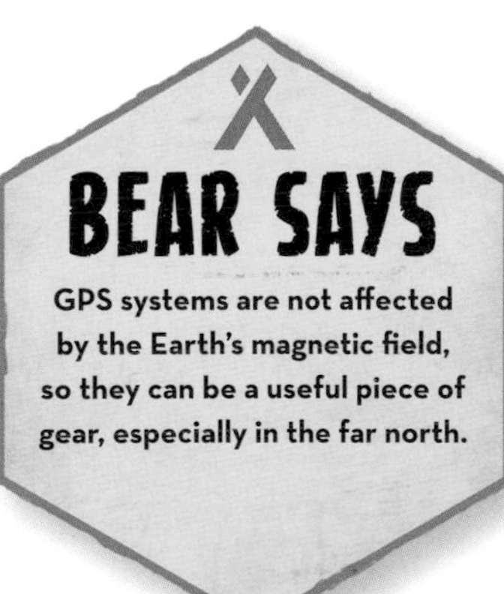

COMPASSES

The compass was invented in China around a thousand years ago and it is still used today by navigators needing to find their direction of travel. It works using a tiny magnet controlled by Earth's magnetic field, which makes the compass point north.

Which compass should I use?

An orienteering compass is a great place to start. It has a transparent base and is designed to be used with a topographical map. It is a very useful piece of gear and, with practice, is easy to use. No matter which way you turn, the magnetic needle always points north, so you can discover which way is north, east, south, and west.

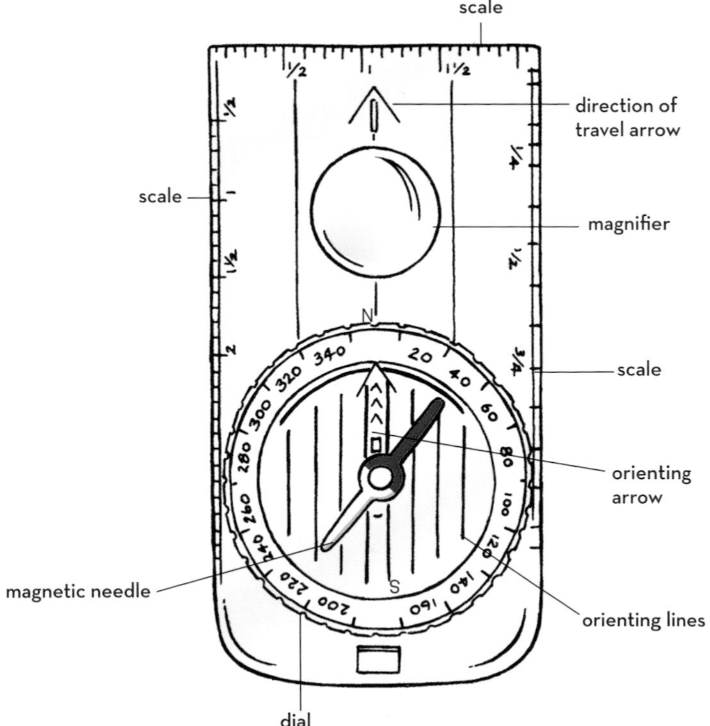

Orient a map

To orient a map you need to align the edge of your compass with the north–south lines of your map. Set the dial of your compass to north, then turn the map and compass together until the north (red) end of the compass needle is directly over the orienting arrow. Your map is now correctly oriented.

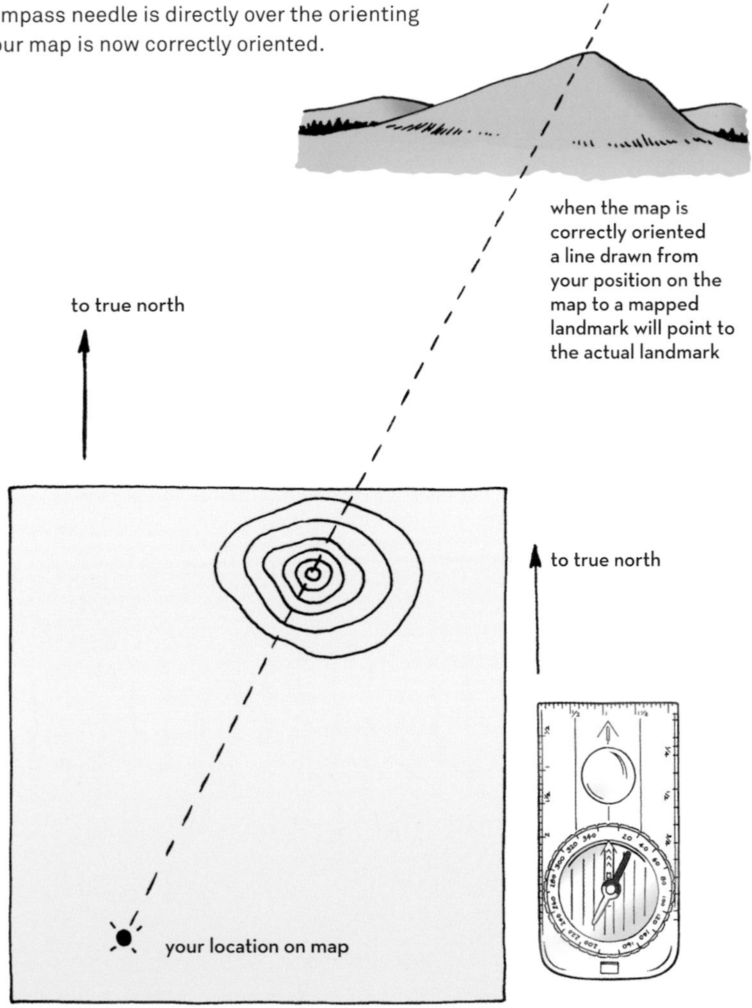

when the map is correctly oriented a line drawn from your position on the map to a mapped landmark will point to the actual landmark

to true north

to true north

your location on map

Take a bearing

The angle between north and an object is called a bearing. They are measured in degrees, for example, 45 degrees (45°). Bearings are always measured clockwise—you start facing north and turn to the right until you reach the angle of the bearing. Then you should be facing your destination. A bearing can also be known as a "magnetic" or "true" bearing depending upon whether it is measured from true north or magnetic north.

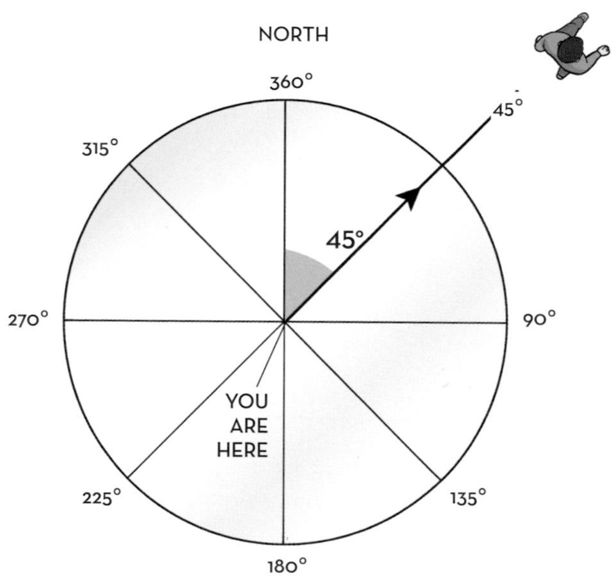

Try it yourself!

These diagrams show a hiker walking in three different directions. Can you work out his bearing for each picture?

BEAR SAYS

Also double check your bearings! If you start out in even slightly the wrong direction you could add hours to your trek.

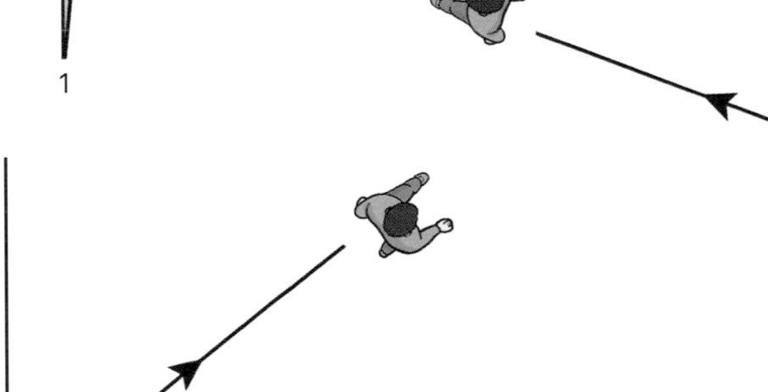

Answers
1. 5° 2. 50° 3. 290°

Get your bearing

1 First align the desired bearing on the dial with the direction of the travel arrow.

2 Hold the compass in front of you and turn your body until the north end of the compass needle is directly over the orienting arrow.

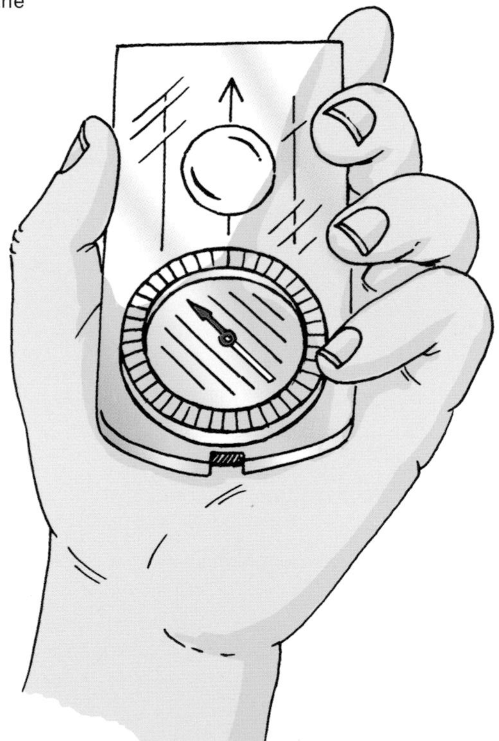

Taking a bearing from a map

Your map does not have to be oriented to take a bearing. Place the compass on the map with the edge lined up against your desired line of travel. Rotate the dial until the "N" is aligned with the top of the map and the orienting lines are parallel with the north–south lines on the map. The bearing on the dial is the bearing you should follow.

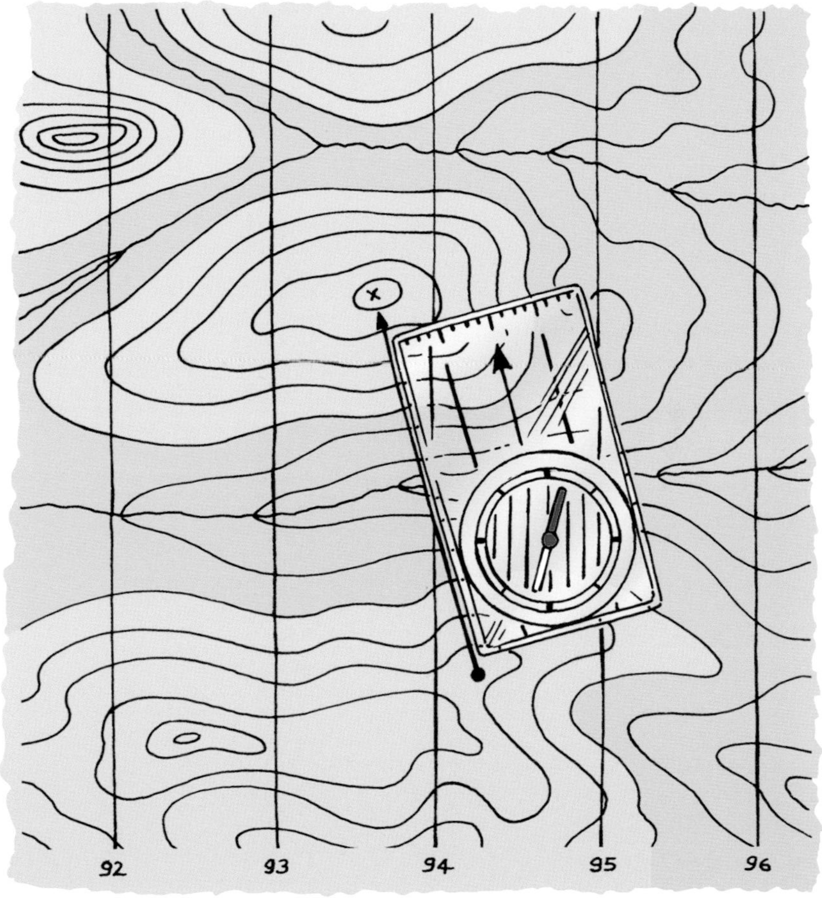

92 93 94 95 96

Bypassing objects

You can also use bearings to make your way around obstacles.

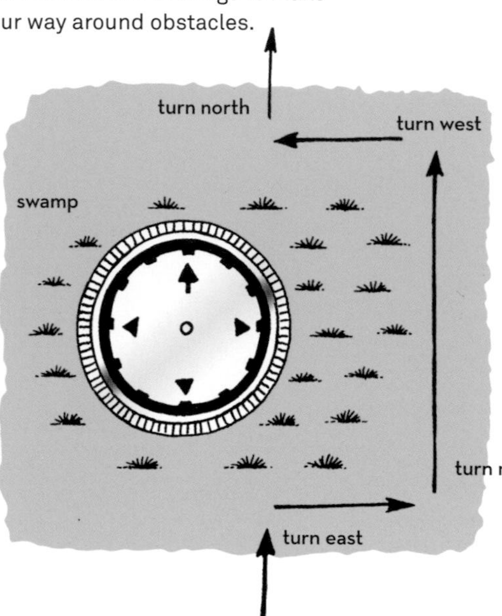

Basic bypass

This bypass involves three 90-degree turns with a fourth to bring you back on track. For accuracy, count your paces when bypassing.

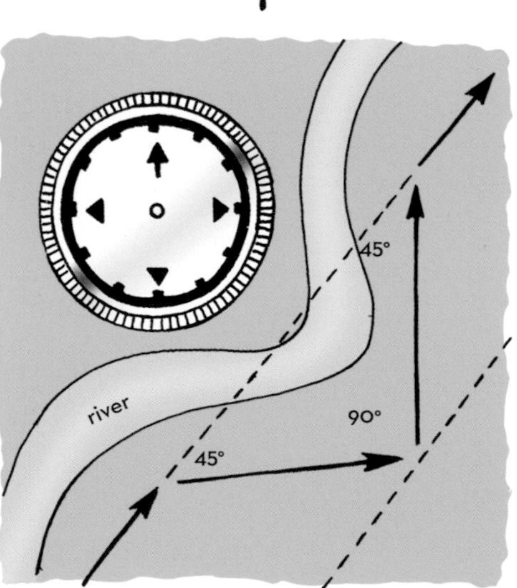

45- and 90-degree bypass

This bypass involves just three turns. Count your paces to ensure that each length is of equal distance.

Marker
Bypassing is easy if you can see a feature beyond the obstacle aligned with your bearing. In this case, walk around the lake until you reach the lone tree.

BEAR SAYS

Learning how to make bypasses to avoid dangerous obstacles can save precious time on a hike that could be spent setting up camp or preparing food.

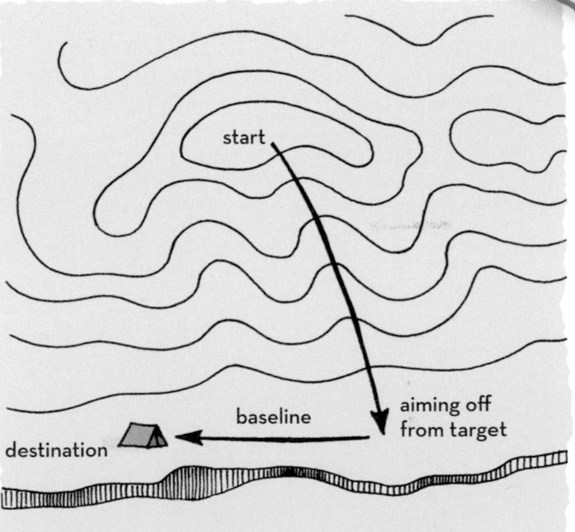

Deliberate offset
Take a bearing left or right of your destination. That way, when you hit a "baseline," such as a road or river, you know which way to turn.

Other types of compasses

Mirror-sighting compass

This compass type is similar to an orienteering compass with the addition of a hinged mirror. This allows you to sight your target and your bearing at the same time.

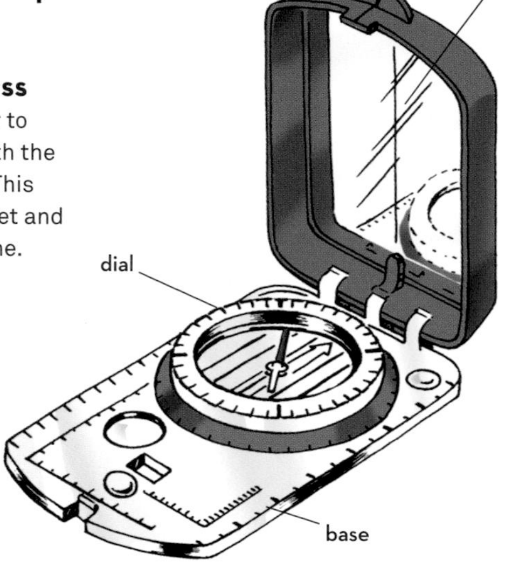

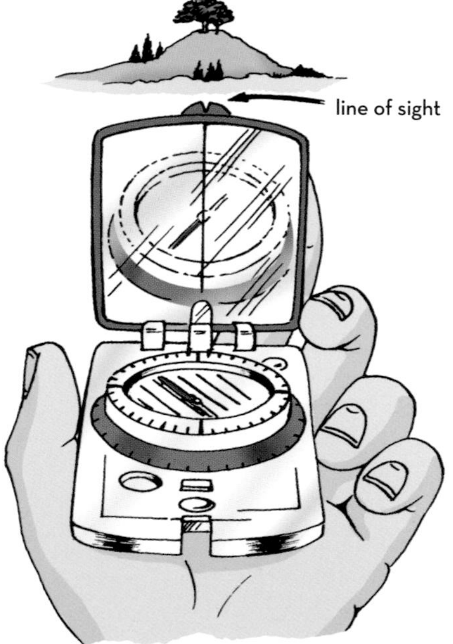

Trying out different compasses is not only useful, but fun too. Each has its own benefits.

Lensatic compass

In situations when an exact reading is required, a lensatic compass is best. They are the favored compass type for military use. Many models have illuminated dial markings so they can be used at night.

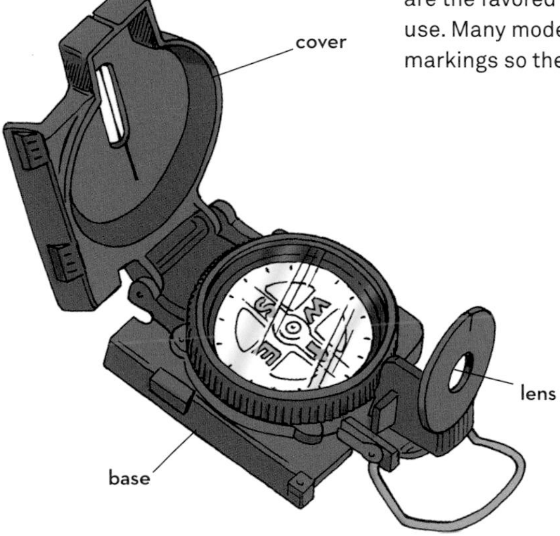

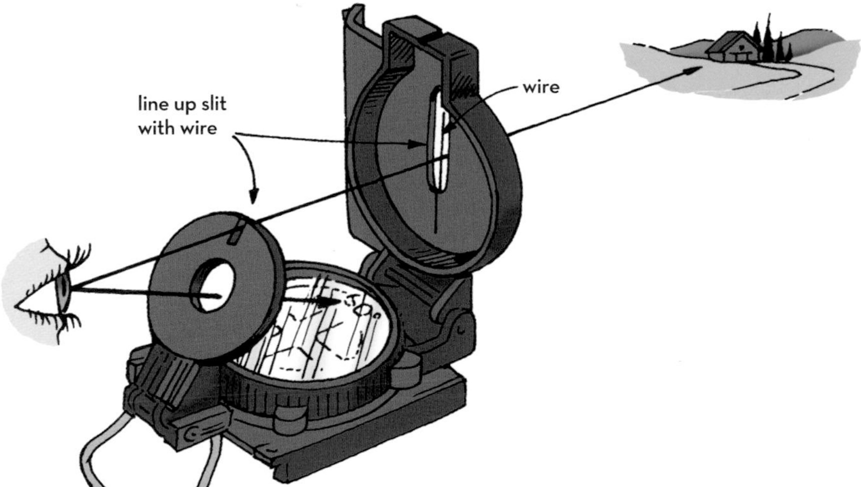

Thumb compass

These thumb-mounted compasses are used in orienteering.

Direct sighting

Look through the eyepiece to get a bearing accurate to within one degree.

Button

A compass doesn't have to be big to be useful. Keep a little one in reserve.

BEAR SAYS

Some mobile phones even come with a compass! You can't rely on the battery life though—so they aren't a replacement for proper gear.

Make your own compass

These are some simple ways to make your own magnetic device.

Magnetize

To make a compass, first magnetize a needle by stroking it in one direction with a magnet. If you stroke toward the point of your needle, the point will indicate north.

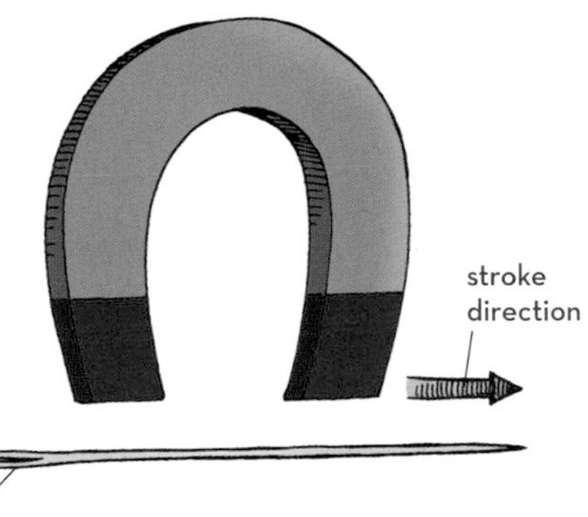

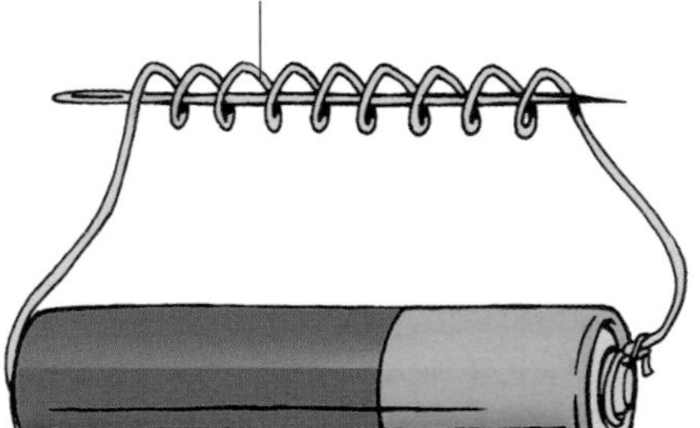

Battery method

Alternatively, you can magnetize a needle by coiling some insulated wire around it. Connect the wire to a battery for five to ten minutes.

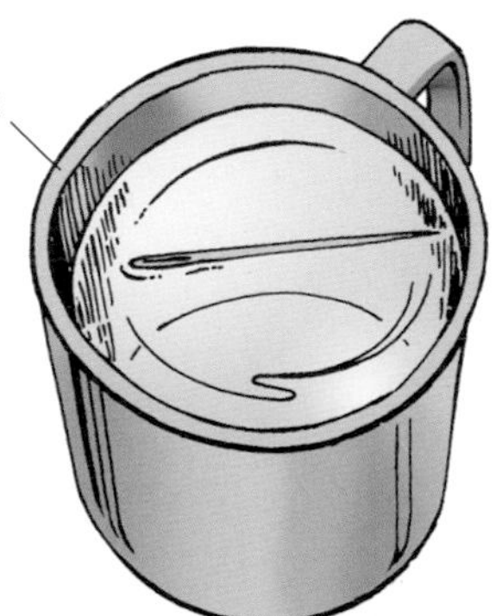

Now you know

If you gently lower the needle into a mug of water, surface tension should keep it floating (a coating of natural oil from your hair will help). The sharp end will point toward magnetic north.

GPS

Hand-held GPS units are a fantastic aid to navigation, and can be the difference between life and death in extreme survival conditions. However, they can't replace maps—you should always carry a compass and know how to use it in case your equipment fails.

The global positioning system

A GPS unit works by measuring its exact distance from a minimum of three satellites in space. The point at which the three signals intersect is shown with a black dot below. This point represents the position of the GPS unit on Earth.

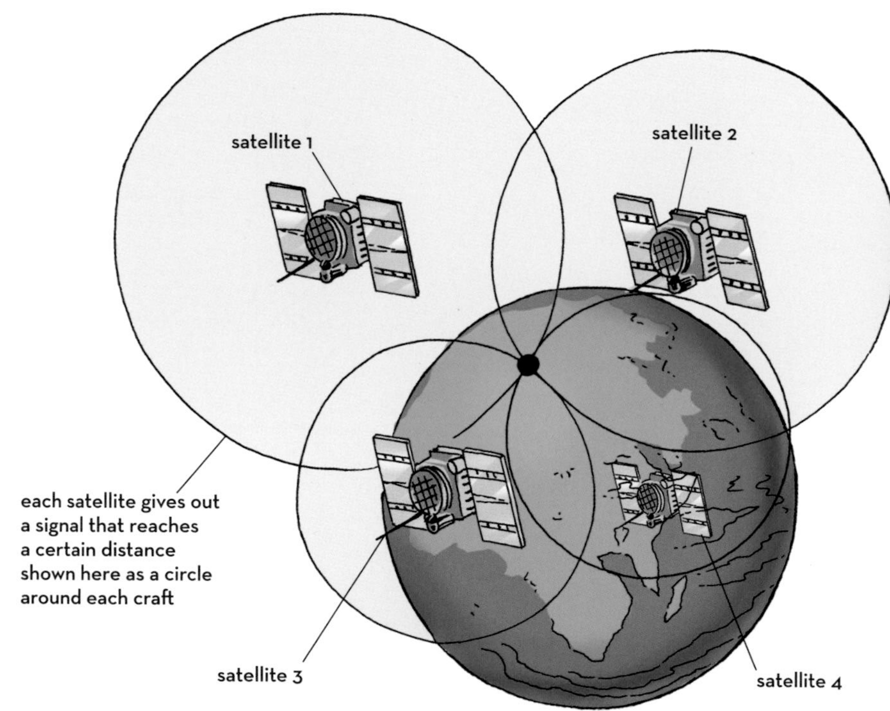

Go on a treasure hunt

If you like exploring, and have a GPS device, you can join the many hundreds of people who enjoy "geocaching." This hobby involves navigating to a set of coordinates and finding a container (the geocache) that has been hidden at that particular spot.

Hand-held unit

A hand-held GPS unit will tell you your location, speed, and approximate altitude (height above sea or ground level). It will also permit you to retrace your path, guide you to specific waypoints or landmarks, and will work as a compass as long as you are moving in one direction. The more expensive models come with built-in maps, electronic compasses, and barometers (devices that measure pressure in the atmosphere).

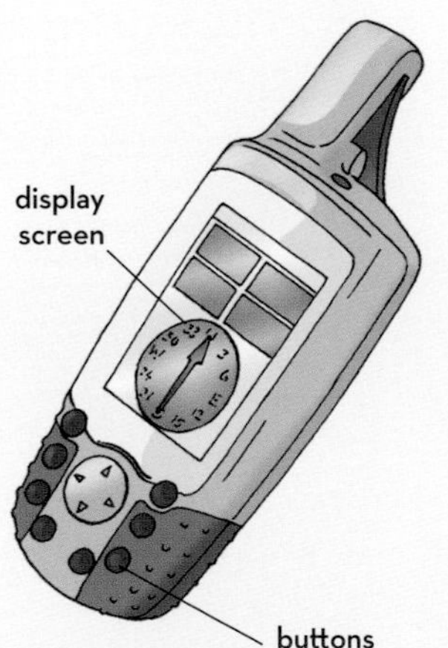

OTHER WAYS TO NAVIGATE

Even without a compass it is possible to determine direction. The sun, moon, and stars can be signposts if you know how to read them. There are also lots of clues in the living world that can help you find your way.

Navigating the night sky

Looking up at the starry night sky is beautiful—and it can be useful too. The North and South Celestial Poles are the points in the night sky that appear to be directly overhead at all times. Finding these points can help you find the right directions. To find the north polar star (Polaris), first locate the pattern of stars known as the Plow or the Big Dipper (Ursa Major). To find the south polar star, first find the constellation Crux.

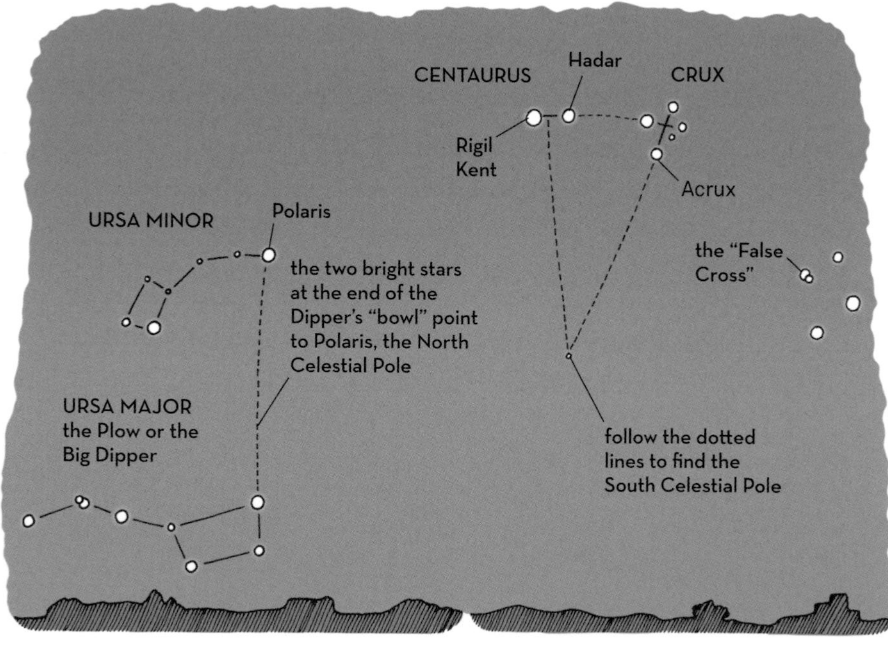

Lunar method

Imagine a line connecting the "horns" of the crescent Moon and project it to the horizon. This point indicates approximate south in the northern hemisphere or north in the southern hemisphere.

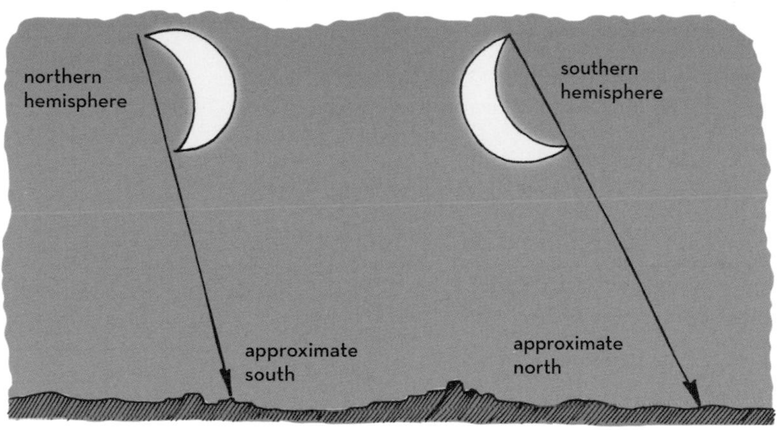

Watch method

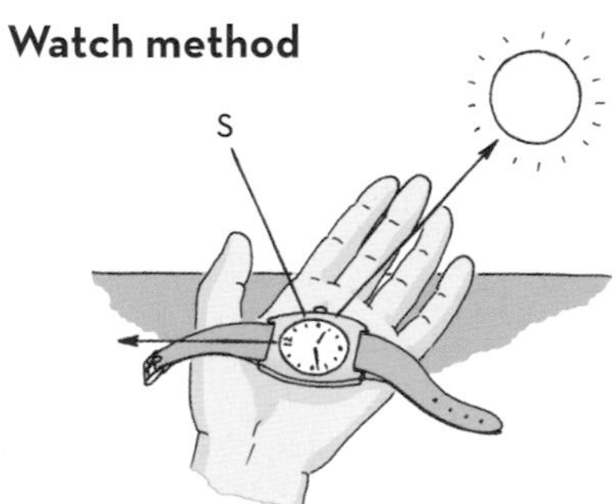

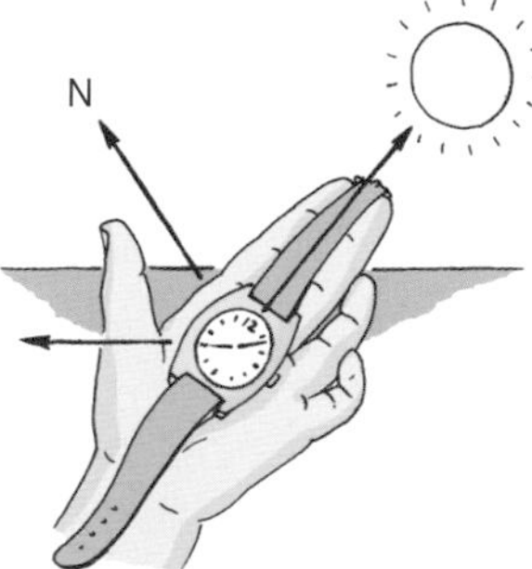

Northern hemisphere

Point the hour hand at the Sun. Divide the angle between the hour hand and the 12 mark in half to find south.

Southern hemisphere

Point the 12 mark at the Sun. Divide the angle between the hour hand and the 12 mark in half to find north.

Shadow tip method

The shadow tip method works in either hemisphere and is most accurate around noon.

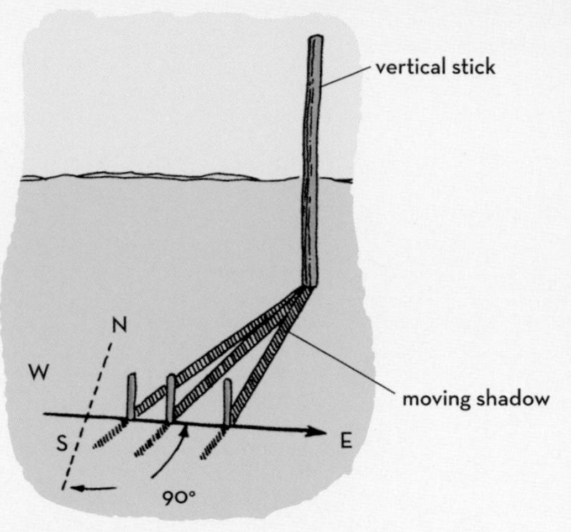

1 First, place a stick vertically in the ground. Mark the tip of the shadow made by the stick with a pebble or another stick.

2 Wait at least ten minutes before marking the shadow's tip with a second marker.

BEAR SAYS

You can always rely on the movement of the sun throughout the day to help pinpoint different directions.

3 As the sun sets in the west the marks will move east. Join the marks to create a line that runs from west to east. Draw another line straight through the east-west line at a 90-degree angle. This will indicate north.

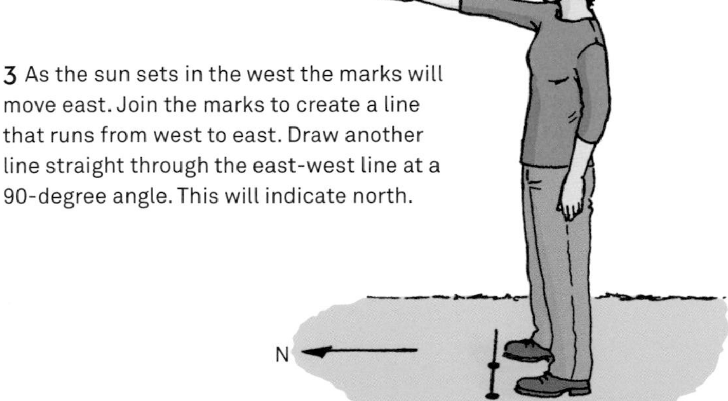

Compass-free map method

Align the features on a map with the corresponding features in the landscape. This will allow you to determine direction and work out your position.

BEAR SAYS

Using landmarks to pinpoint your location is something you can practice easily. Get a local map and go on a walk with a parent to give it a try.

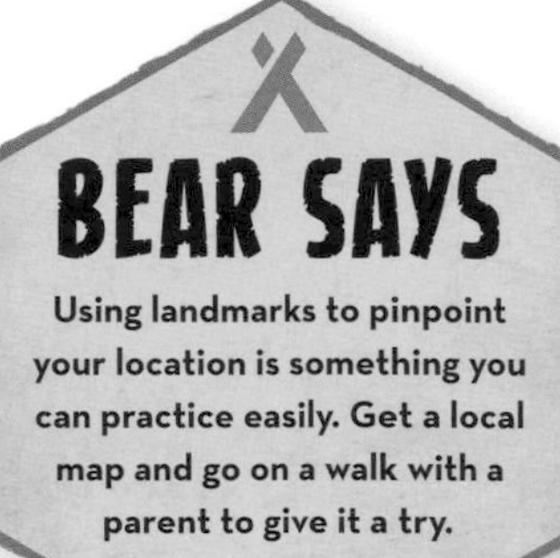

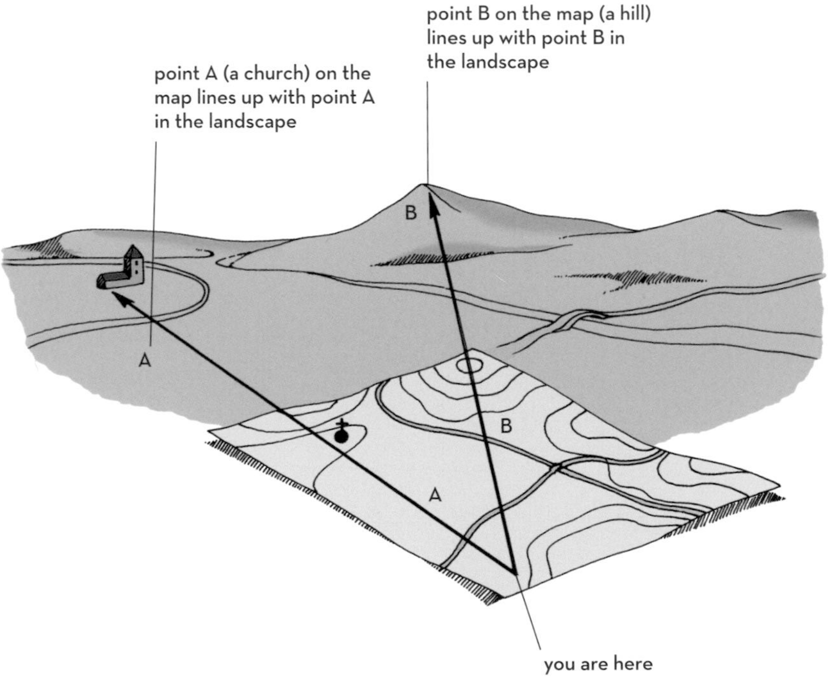

point B on the map (a hill) lines up with point B in the landscape
point A (a church) on the map lines up with point A in the landscape
B
A
B
A
you are here

Night star method

This simple method uses stars to pinpoint your location.

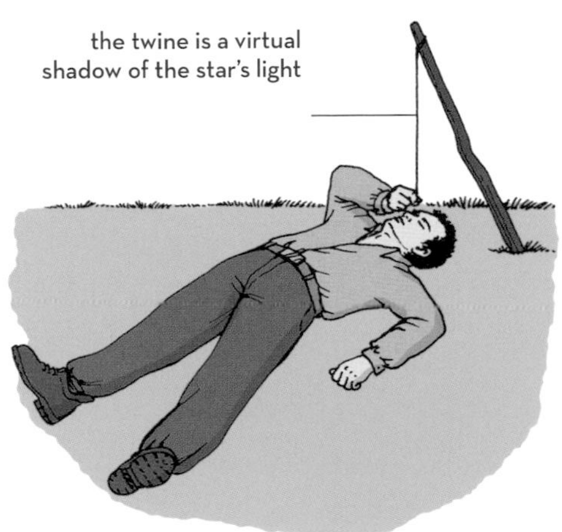

1 Set a long stick into the ground at a slight angle. Tie a length of twine to the end of the stick. Lie on your back, place the twine by your temple, and align it with a well-known star.

2 Mark the star's position on the ground and then wait a while before marking it for a second time. As with the shadow tip method, the first mark will be west and the second will be east.

mark the position with a stick

Signs in nature

Snow

Melted snow on one side of a tree will indicate south in the northern hemisphere.

Glacial boulders

These large rocks sit on pedestals of ice that erode on the south side in the northern hemisphere.

Green sign

Moss usually grows better on the shady side of a tree trunk – that's the north side in the northern hemisphere.

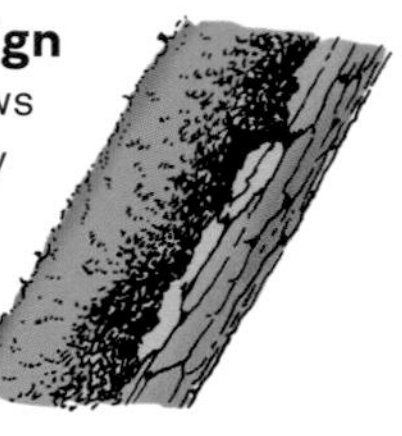

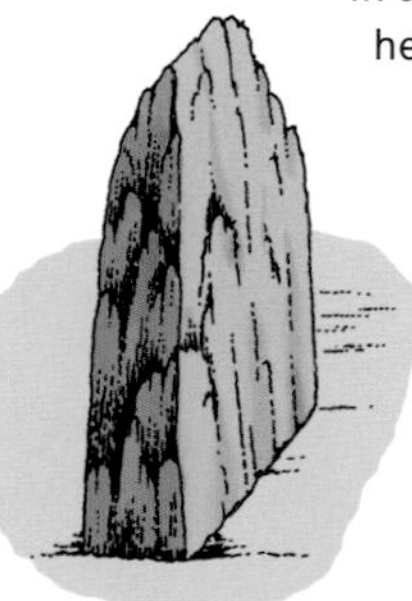

Magnetic termites

These Australian termites build their mounds aligned north–south to minimize exposure to the sun.

Traveler's palm

The leaves of this palm form a dramatic fan that is usually aligned east–west.

Nesting birds

If you are near water, look for signs of breeding wildlife, e.g. nests or frogspawn. Water birds, fish, and frogs often breed on the west side of a river or lake.

Prevailing wind

The most common wind direction that a location experiences is called the prevailing wind. The place the wind has come from brings different types of weather. The prevailing south-west wind in Britain, for example, is partly why it often rains more than in other countries.

Spider webs

Do you know which way the wind tends to blow? Spiders do, and will orient their webs sideways to the prevailing wind.

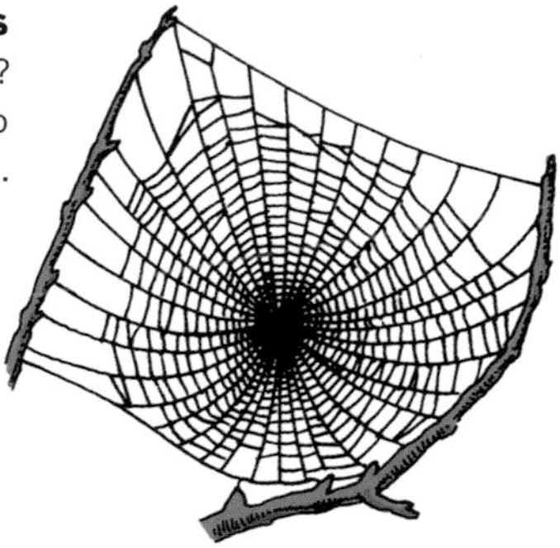

BEAR SAYS

When all else fails, turn to nature for clues to your surroundings.

Trees

Windswept trees are another good indicator of prevailing winds, and will show you the way if you know the local weather.

KNOTS

People have used the knots included in this book for years and they are just as valuable today. Early uses included making shelters, weaving, fishing, and tethering animals. Although they can take a while to master, the moment you tie a tricky knot and get to use it in action, you realize how exciting this ancient art can be!

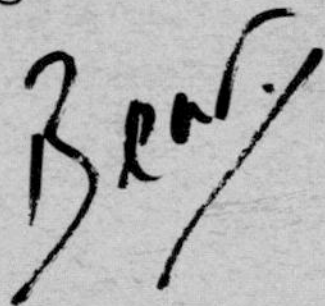

IN THIS SECTION:

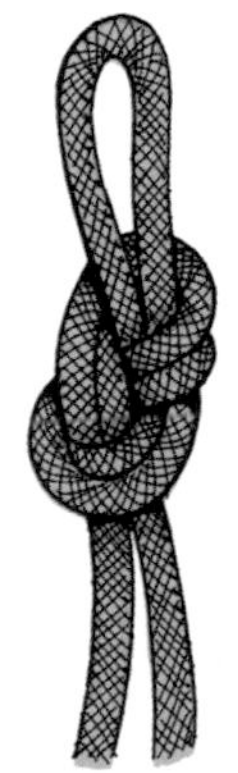

DISCOVER KNOTS!

A sound knowledge of ropes and knots is extremely important for climbers and mountaineers, but they also have uses in everyday life. It takes time and practice to master the art of knot tying, but it is very rewarding to put your skills to use!

Types of rope

Static ropes do not stretch, while dynamic ropes allow some movement if the load they are holding suddenly falls.

Laid rope

This rope is used in many outdoor activities. It has three or more strands twisted around each other.

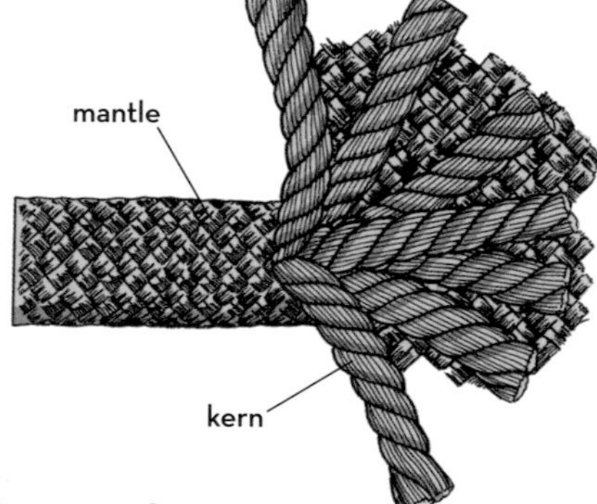

Kernmantle rope

Many climbing ropes have a protective sheath, called a mantle, covering the core fibers (kern).

Damaged ropes

The sheath of kernmantle ropes can hide a damaged core. Bulges in the core or tears in the mantle mean that the rope needs to be replaced.

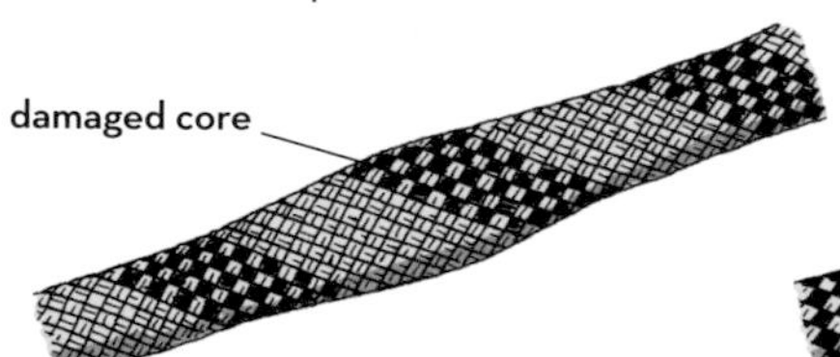

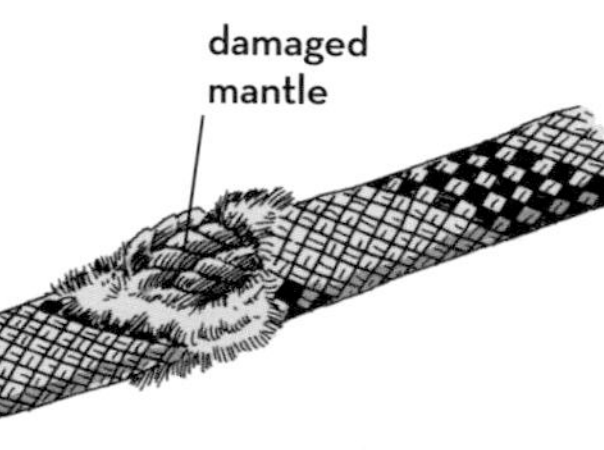

Learn the terms

Turn	The U-shape that is formed when a rope is hung over a rail.
Round turn	A round turn is created when a rope is hung over a rail, and then wound around the rail one more time.
Half hitch	This knot is made when a rope is hung over a rail, then either under, or both under and over, itself.
Clove hitch	Two half hitches that are connected and placed next to each other make a clove hitch.
Standing end	The standing end, or standing part, refers to the main body of the rope.
Overhand turn	A circular shape in the rope where the working end finishes on top of the circle.
Underhand	A circular loop in the rope where the working end sits on the underside.
Working end	The portion of rope, other than the standing part, that is used to tie a knot. When the knot is finished, the working end is usually the small tail left over.
Bight	A bight is a U-shaped form made in the rope as part of the knot-tying process.

Carrying ropes

1 Put the two rope ends together. Measure out two arm lengths, then loop the rest of the rope in double arm lengths over your knee or around your shoulders. This is the main section that will sit on your back.

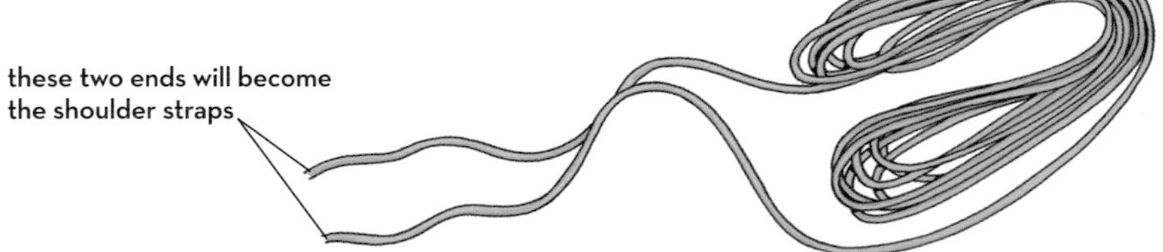

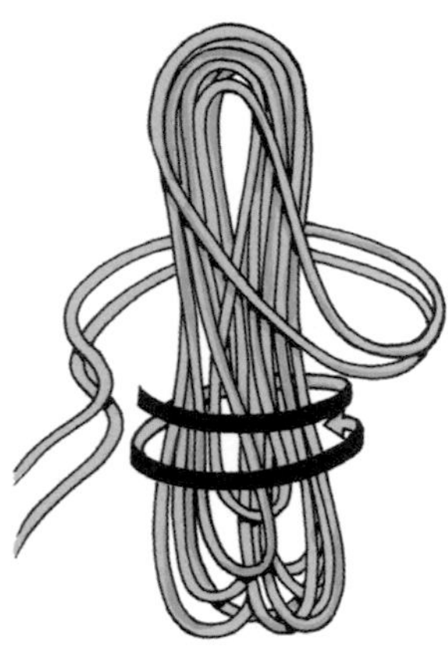

2 Wrap the two long ends tightly around the whole coil several times.

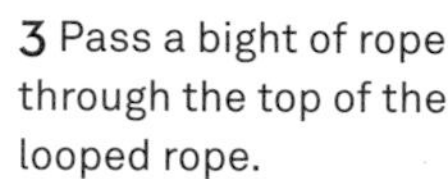

3 Pass a bight of rope through the top of the looped rope.

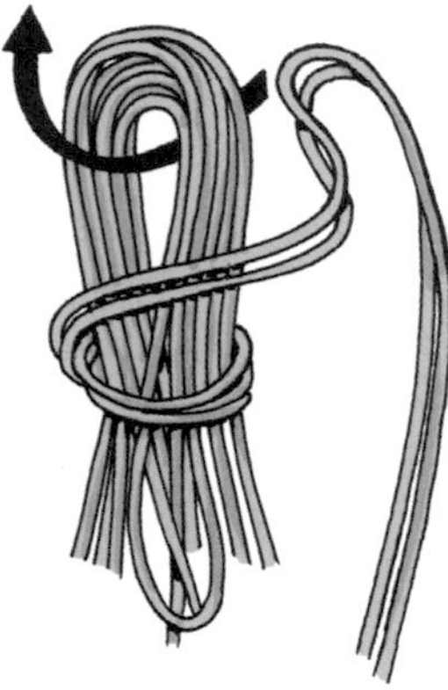

4 Run the loose ends through the bight of rope.

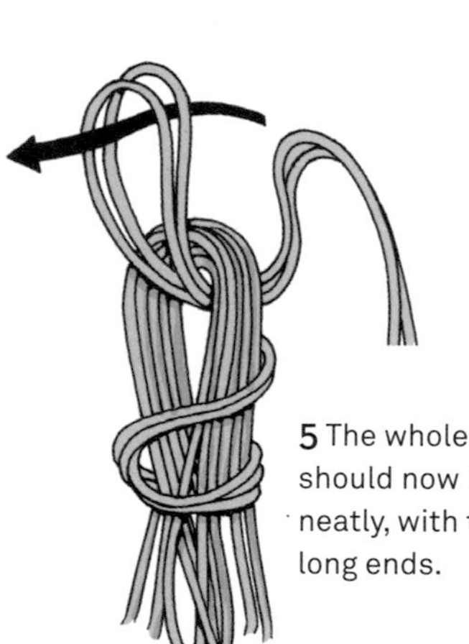

5 The whole rope should now hang neatly, with two long ends.

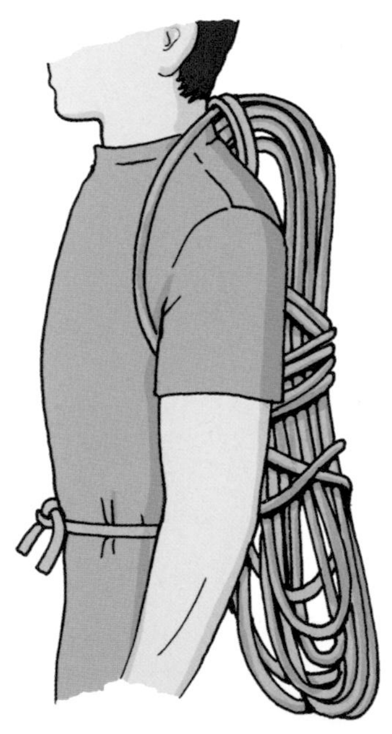

6 Run the long ends over your shoulders, then across your back, and tie across your stomach.

OVERHAND LOOP

One of the quickest ways to make a fixed loop in a rope is the overhand loop, suitable for a quick secure on a belay station. This knot can become very tight, however, and can be difficult to undo, so it isn't always the best choice. This is sometimes called a "thumb knot."

Loop knots are closed bights that have been tied either in the bight itself, or at the end of the rope.

1 Put the two rope ends together. Stretch the loop out over the top of the trailing rope.

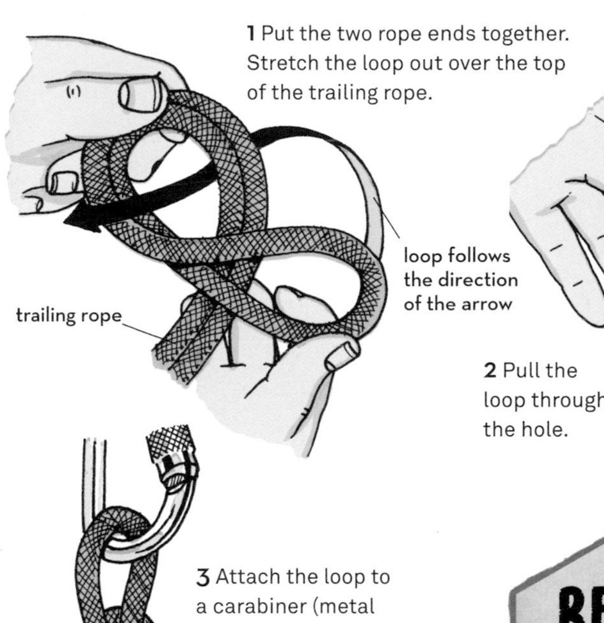

2 Pull the loop through the hole.

3 Attach the loop to a carabiner (metal loop, with a spring-loaded opening).

BEAR SAYS

This knot is suitable for a quick secure on a belay station (the point on the rope from which a climber can hang and lower their partner down).

FIGURE-EIGHT LOOP

Although a bulkier knot, this is thought to be one of the most secure ways to create a loop in a rope, and has many uses.

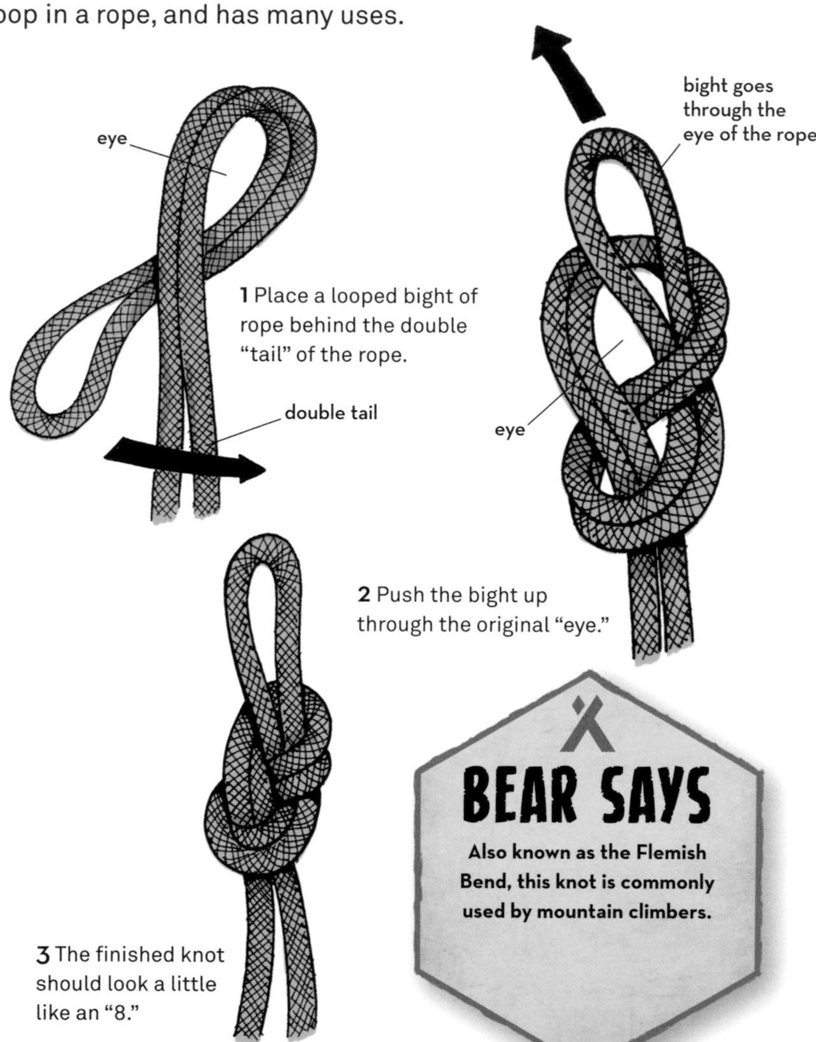

1 Place a looped bight of rope behind the double "tail" of the rope.

2 Push the bight up through the original "eye."

3 The finished knot should look a little like an "8."

BEAR SAYS

Also known as the Flemish Bend, this knot is commonly used by mountain climbers.

THREADED FIGURE EIGHT

This should end up looking the same as the figure-eight loop knot, but it allows you to tie in to a fixed point, such as a harness or sling. It is done in two parts.

1 Start by tying a single, loose figure-eight knot in the rope, leaving a long tail. Thread the tail through the harness or sling, then start threading the rope back through the knot.

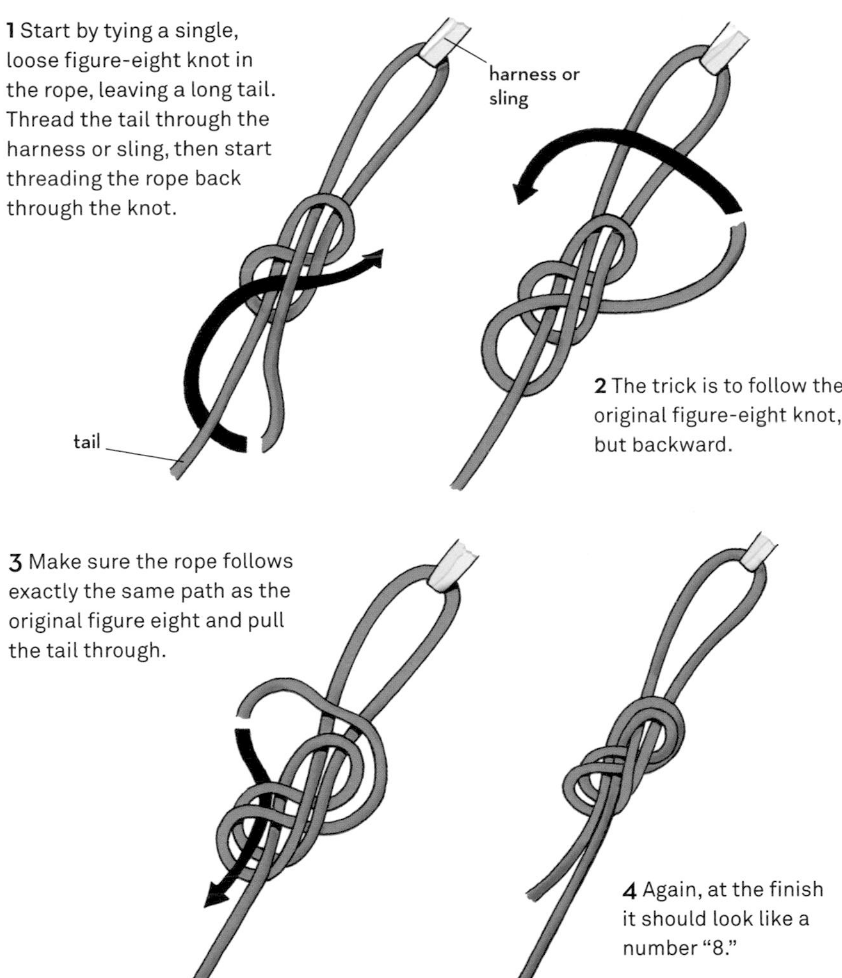

2 The trick is to follow the original figure-eight knot, but backward.

3 Make sure the rope follows exactly the same path as the original figure eight and pull the tail through.

4 Again, at the finish it should look like a number "8."

JURY MAST KNOT

This knot can be used to "jury rig" (temporarily fix) a boat's mast if the existing rigging fails. The ropes holding the makeshift mast upright attach to the knot's three adjustable loops. A jury mast knot can also be used to put up a tent or flagpole, as long as there is something to prevent the knot sliding down.

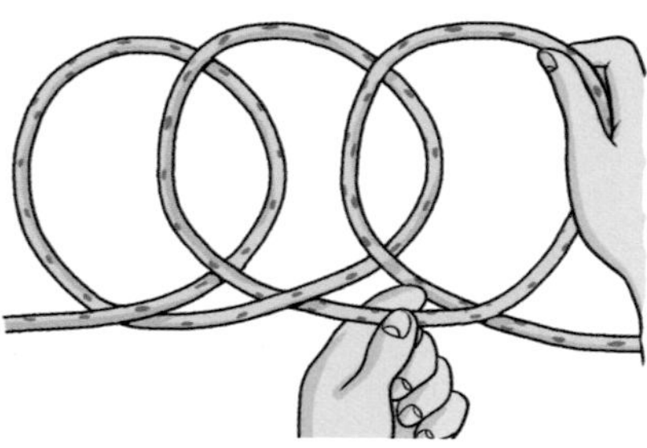

1 Working from left to right, make three underhand turns. The left edges of the second and third turns should overlap the right edges of the first and second turns.

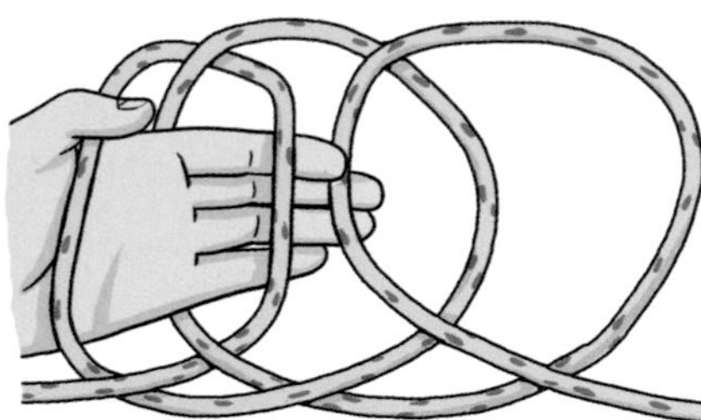

2 Lead the left hand into the first turn from underneath, over the left edge of the second turn, under the right edge of the first turn, and pick up the left edge of the third.

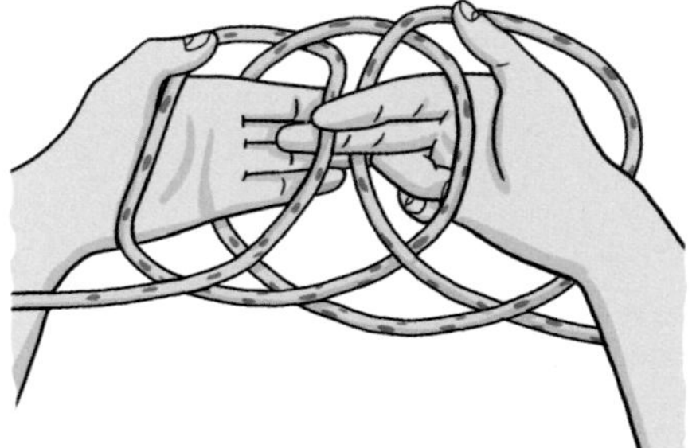

3 With the right hand, lead over the right edge of the third turn, under the right edge of the second turn, over the left edge of the third turn, and pick up the right edge of the first.

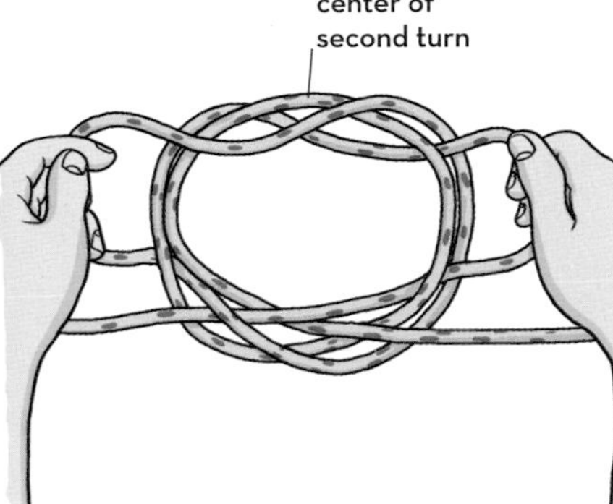

4 Draw your hands apart far enough so that two loose bights begin to form. The remainder of the knot should be circular in shape with an obvious hole in the middle.

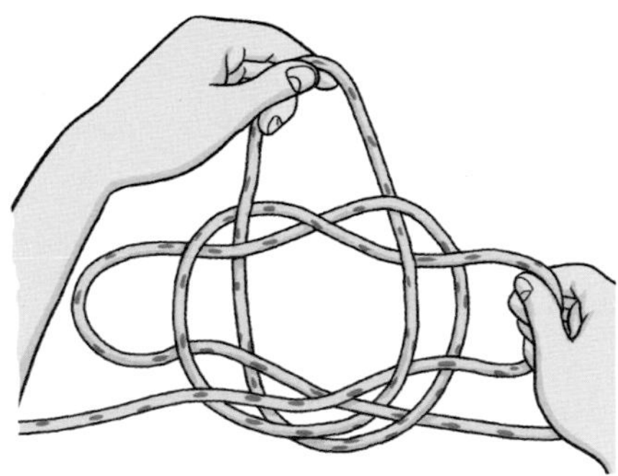

5 The second turn is still in the circular portion of the knot. Draw it out carefully at the top to form a third bight. Adjust the knot so that the three bights are of a similar size.

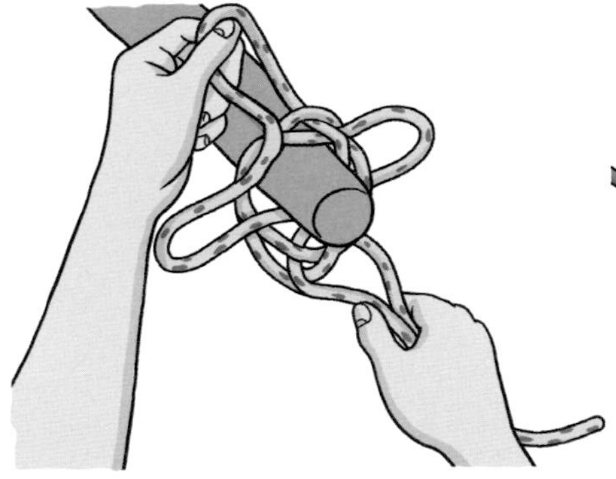

6 Place the knot over the end of a spar, and adjust to fit snugly. Tie both standing ends together around the spar. The three bights become attachment points for stays.

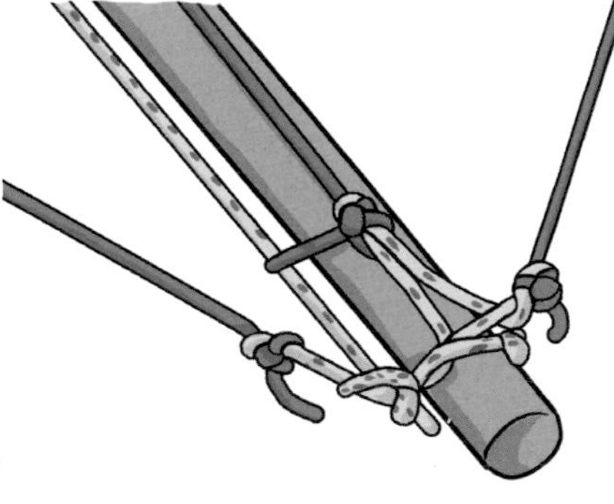

7 The stays—ropes or wires supporting the mast—can be attached to the jury mast knot with sheet bends or double sheet bends.

PERFECTION LOOP

The perfection loop is a type of knot that forms a fixed, single loop, normally tied at the end of a rope. The perfection loop is difficult or impossible to untie, so it is best used where the line can be cut once the loop is no longer needed.

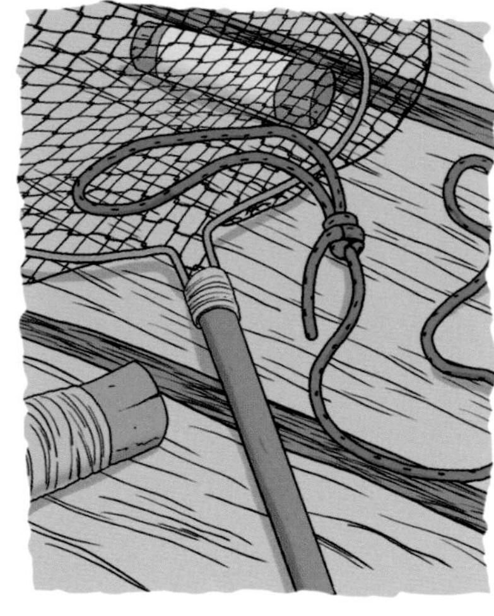

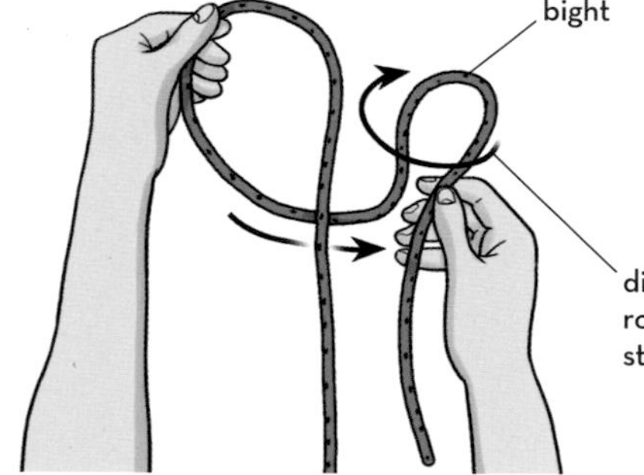

1 Begin with a working part twice the length of the loop you wish to make. Make an underhand turn and then form a bight in the working end.

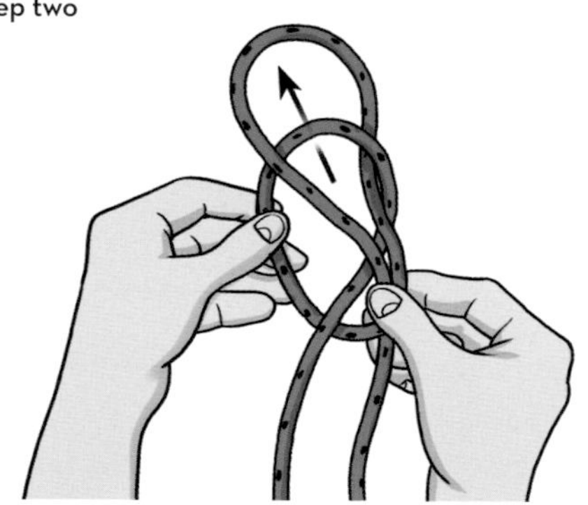

2 Rotate the bight right to left, then lead it through the turn from front to back. The bight will become the loop.

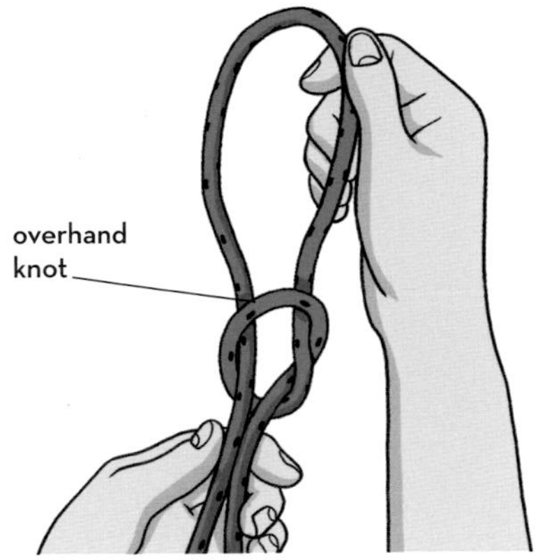

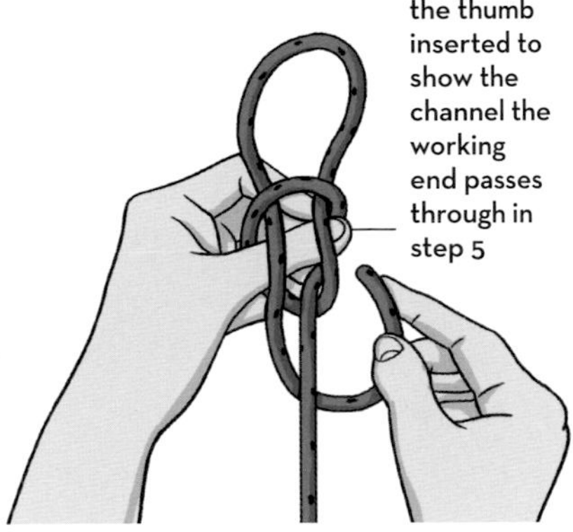

3 If the working end is too long, pull the extra cord into the loop. Work it out through the overhand knot and into the standing end.

4 Lead the working end around behind the standing part, next to the lower portion of the overhand knot.

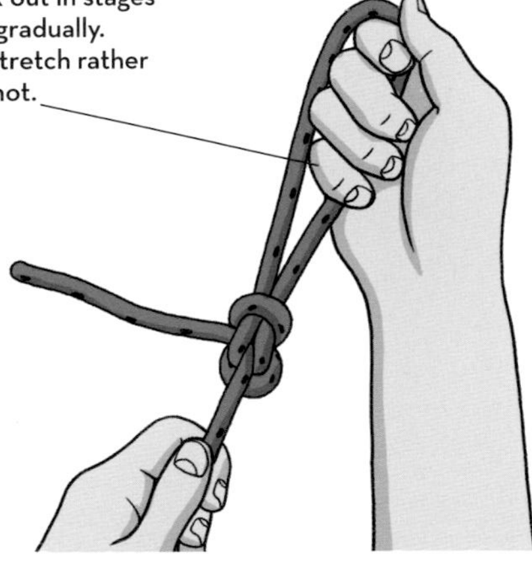

5 Slide the working end under both strands of the loop, from right to left.

6 Tighten the knot by pulling on the loop, the standing part, and the working end. Pull hard on the loop and standing part to complete, particularly in shock cord, which stretches.

BOTTLE SLING

This knot provides a carry handle for a bottle. It can be used to cool drinks in cold water slung over the side of a boat. This example requires about 5 ft. of cord.

Binding knots are tied in a single piece of rope, around bundles of objects.

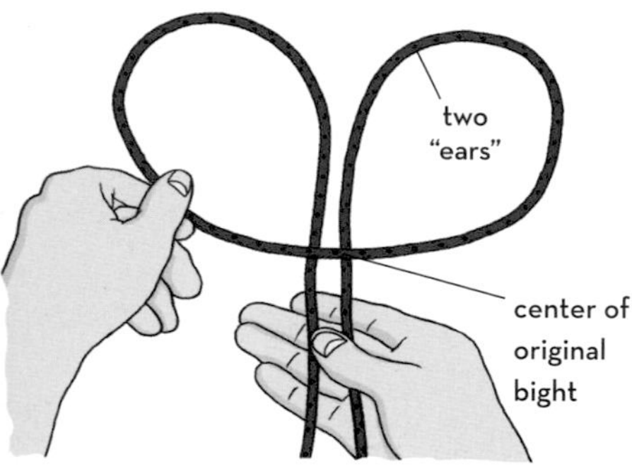

1 Fold the rope in half, and place the bight flat on the table. Fold the bight down as shown, to make two even-sized "ears."

2 Slightly overlap the right ear over the left. The center of the original bight should still be below the intersection of the two ears.

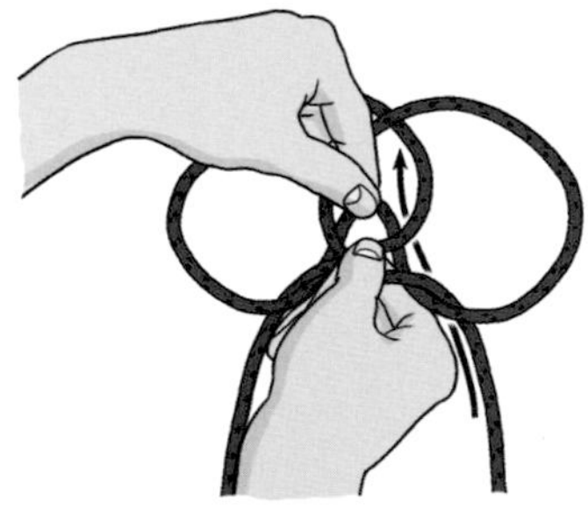

3 Holding the pattern in place on the table, draw the center of the original bight under the knot at the point of the lower intersection of the ears.

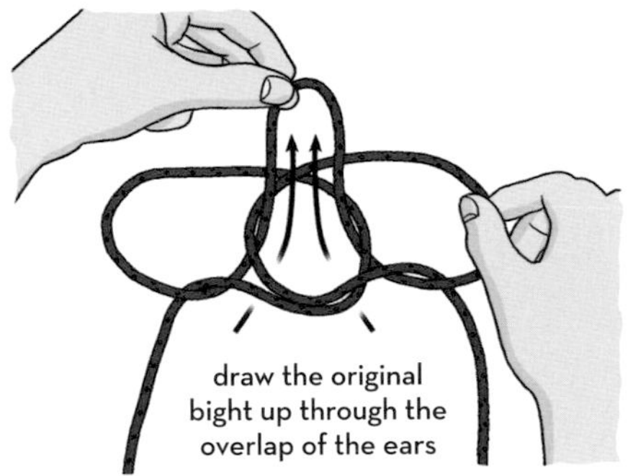

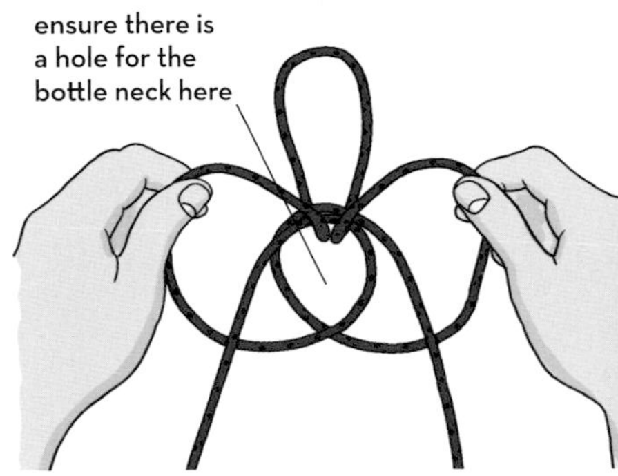

4 Now bring the center of the original bight up through the space formed by the overlapping ears, to make a new bight at the top.

5 Using both hands, turn the ears and center part over at the same time, with the top moving away from you. The top of the ears now finish at the bottom of the pattern.

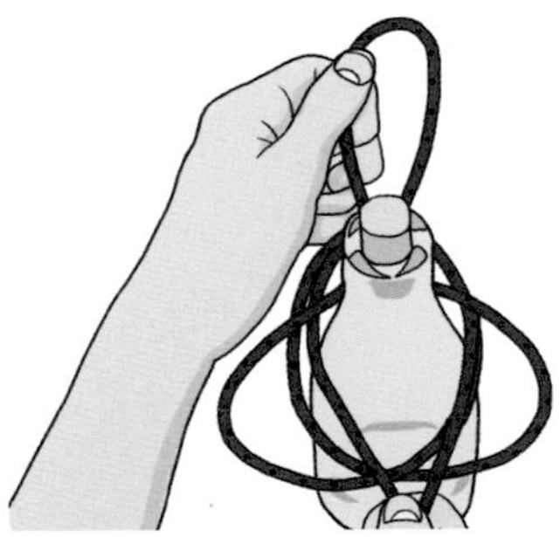

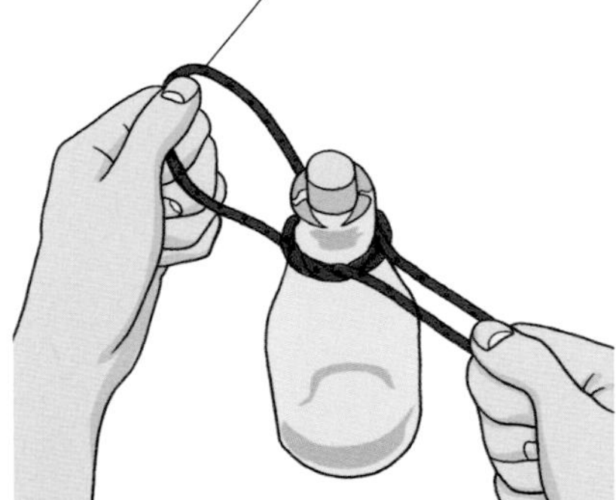

6 Lift the knot carefully and place the hole over the neck of the bottle. Tighten by pulling on the upper bight and evenly on the pair of ends.

7 Work the knot to a snug fit around the neck of the bottle. The ends can be tied together to make a second handle, using a fisherman's knot or figure-eight bend.

ANCHOR BEND

The anchor bend, also known as the fisherman's bend, is quite a secure knot. It is most often finished by the addition of a half hitch. It will hold in bungee cord, whereas the "round turn and two half hitches" is particularly simple to tie, but it will not hold at all in bungee cord.

Bends are types of knots that join two separate pieces of rope.

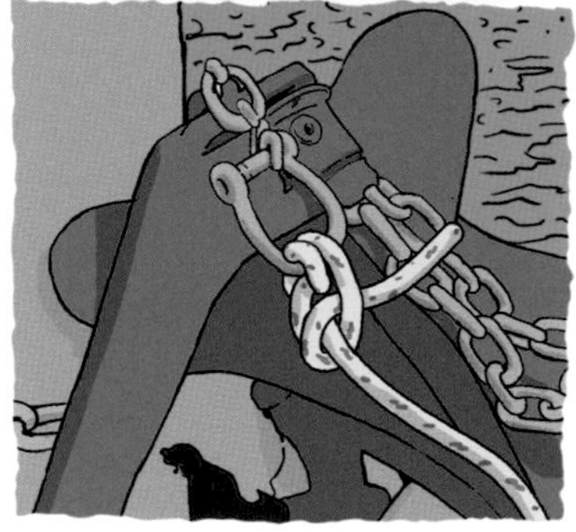

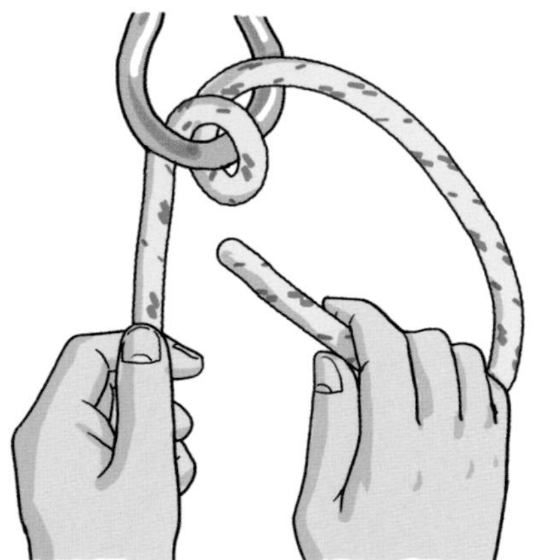

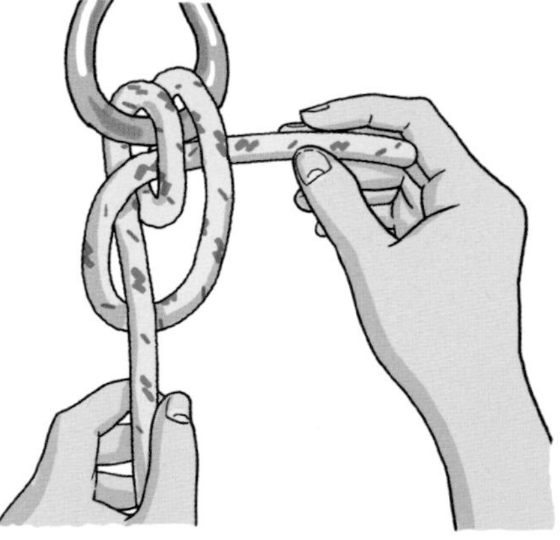

1 Make a round turn through a ring, winding from back to front and left to right. Do not pull the turn tight yet, as the working end needs to pass through it first.

2 Lead the working end left behind the standing part, then to the front, and tuck it to the right beneath both windings of the round turn.

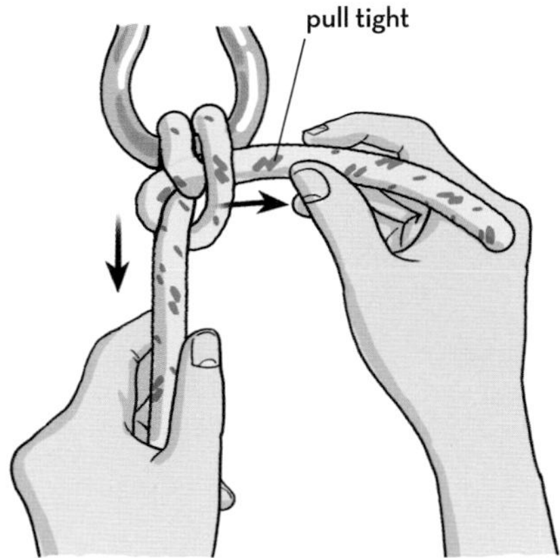

3 Pull on both the standing part and working end to tighten the knot around the ring. This is a completed anchor bend, but a half hitch is usually added.

4 Pass the working end around behind the standing part again, to the front, and under itself. Pull the working end as tight as possible.

Round turn and two half hitches

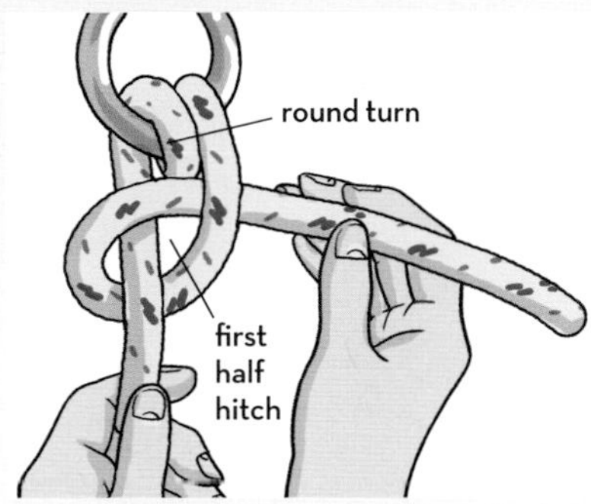

1 Make a round turn through the ring as in step one. Pass the working end around behind the standing part, to the front, and under itself, to make the first half hitch. Pull tight.

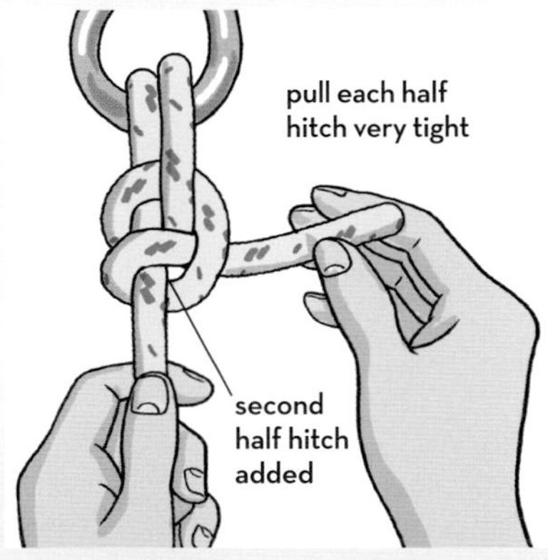

2 Continue the working end in the same direction, around behind the standing part and to the front again, and make the second half hitch.

DOUBLE FISHERMAN'S KNOT

This is one of the best knots to safely tie two ropes together, and is useful on long descents. Start by laying the last 3 ft. of each rope alongside each other, tails in opposite directions.

1 Turn the end of one rope around the other rope twice, and pass the end back through the loops, away from the knot's center. Do this to both ropes.

2 This should form two "x" shapes, which will then slide together as you pull each rope.

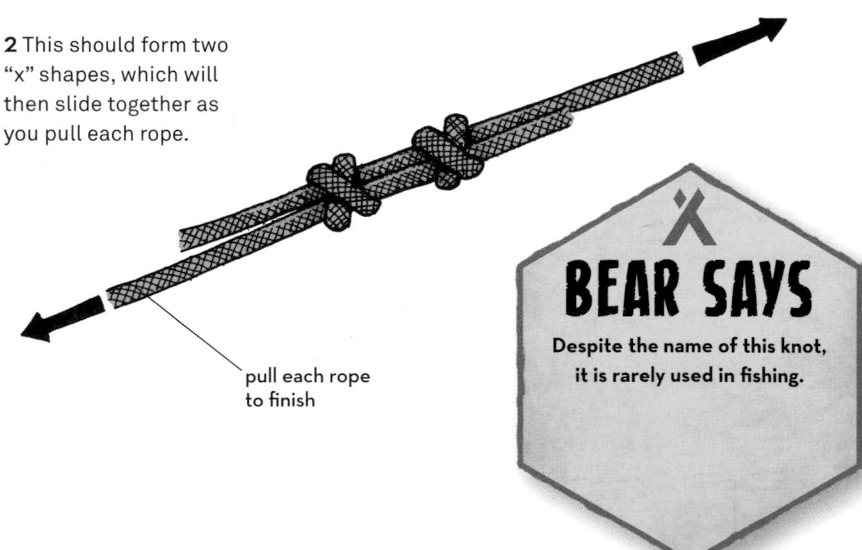

BEAR SAYS

Despite the name of this knot, it is rarely used in fishing.

PRUSIK HITCH

The prusik knot is an excellent way to attach a weight to a rope. It slides up and down the rope when unweighted, but doesn't slip under a downward force. Two prusik knots (one for your feet, one clipped into the harness) are often used to climb up, or "prusik," a rope.

Hitches are used to attach load-bearing cordage to objects. The choice of hitch is important if the knot needs to remain secure.

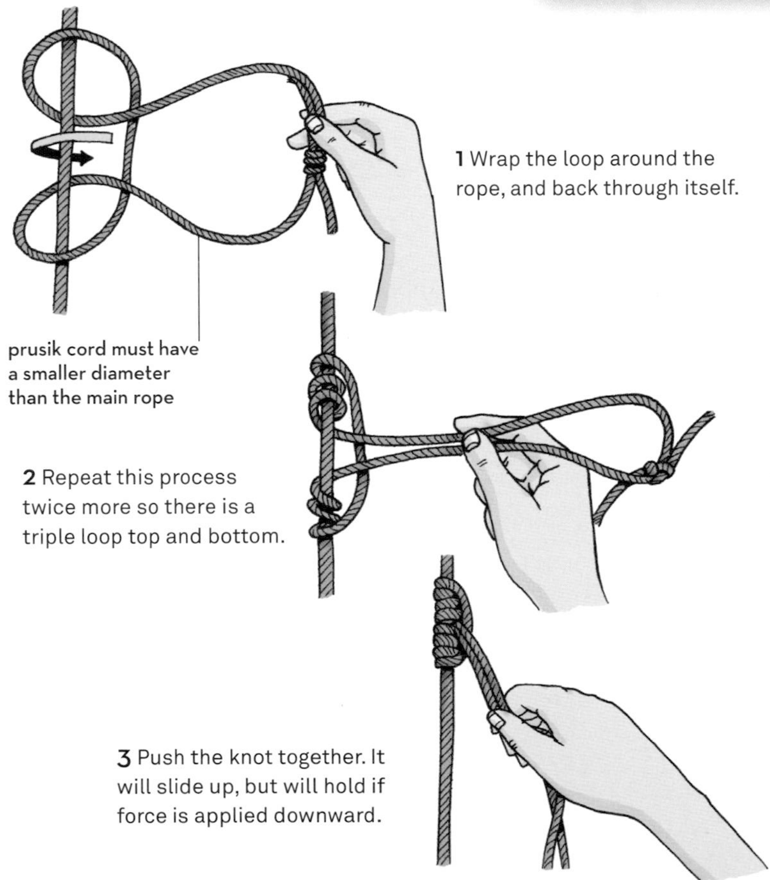

1 Wrap the loop around the rope, and back through itself.

2 Repeat this process twice more so there is a triple loop top and bottom.

3 Push the knot together. It will slide up, but will hold if force is applied downward.

CLOVE HITCHES

The clove hitch is a quick and memorable knot that can be used to attach a rope to a pole or a carabiner. It may slip on a smooth surface. The load can be applied to either end of the rope. Steps one and two show the clove hitch tied with a working end. The alternate "in-the-hand" method allows the hitch to be tied in the bight—anywhere along the rope.

1 Lead the working end over and down behind the pole, then up in front and over itself to the left.

2 Lead the working end diagonally across the turn, around the back of the rail again, then upward under itself. You have formed two half hitches. Pull the ends to tighten.

Alternate clove hitch

This method can be used when you can slip the hitch over the end of a pole.

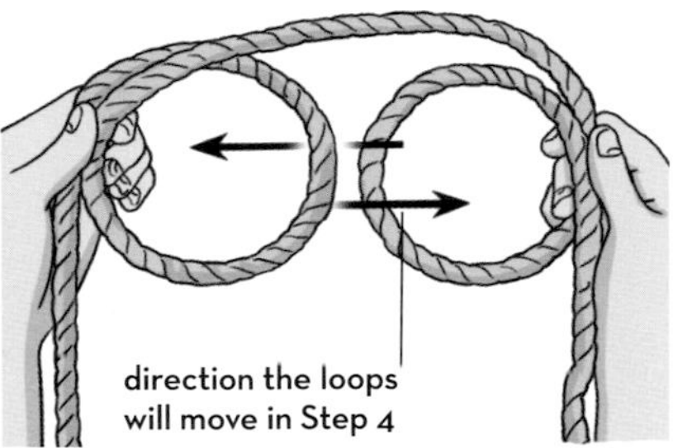

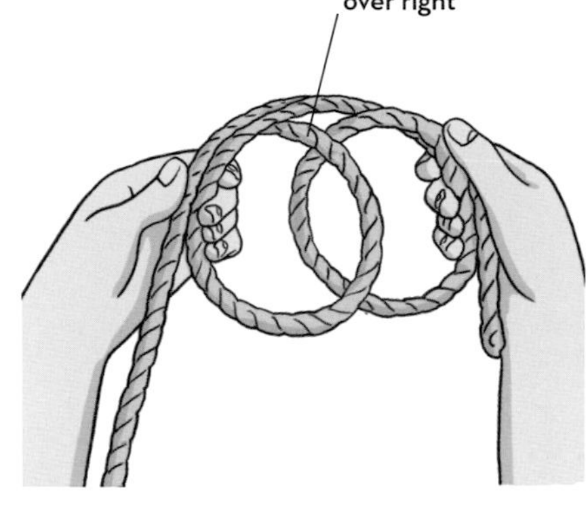

1 If you don't have easy access to the ends of the rope, tie the clove hitch using this alternate method. Begin by making two consecutive overhand turns.

2 Slide the left-hand turn over the right, then slide the fingers of the left hand through the center of both loops.

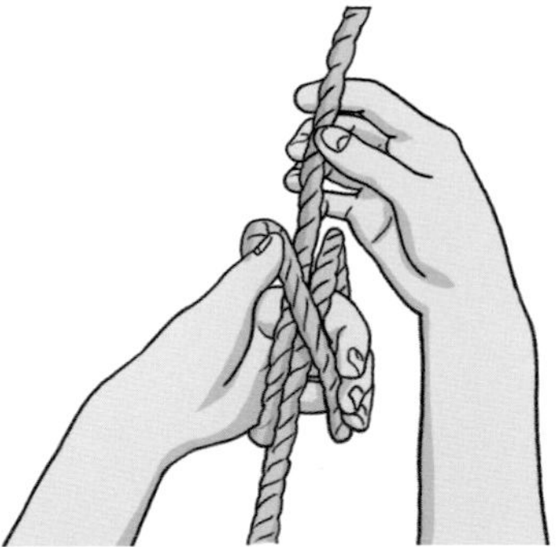

3 When viewed from the side, the two half hitches can be identified. It is the same knot as shown in step two of the simpler clove hitch, shown to the left.

4 Simply slide the half hitches over the end of the pole and tighten.

Clove hitch for a carabiner

This hitch is also used in climbing. It can be used to safely lower climbing gear from a great height. It increases the amount of friction on the rope but also allows it to slip, so it can be used instead of a belay.

The clove hitch is also known as a double hitch. It is one of the most commonly used in the scouts, so give it a try!

1 Make two loops in the rope, both in the same direction.

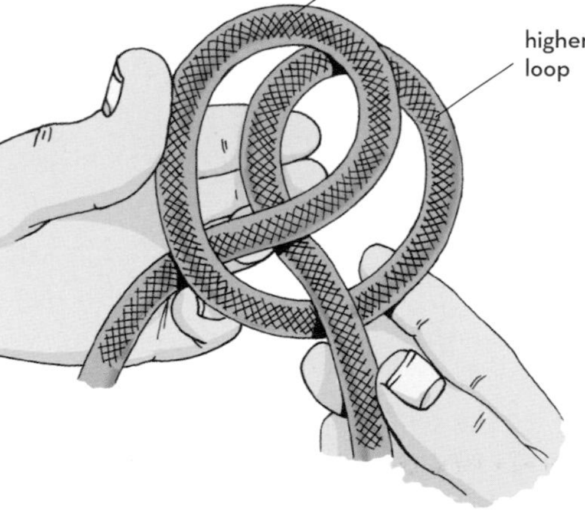

2 Place the lower loop over the higher loop, so that both tails of the rope are in the middle of the hitch.

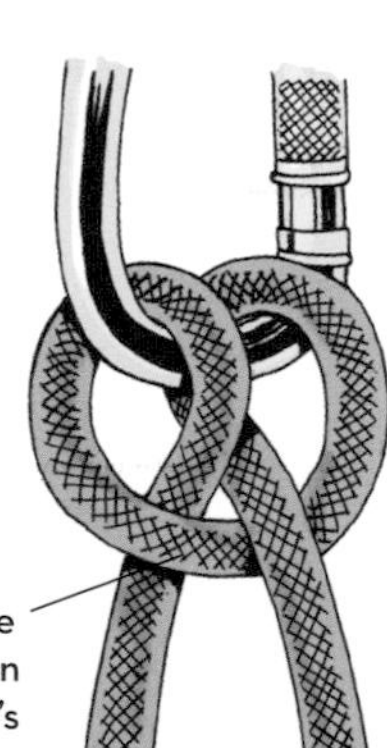

3 Clip a carabiner through the two loops. The hitch is surprisingly strong.

ITALIAN HITCH

This hitch can be used to safely lower climbing gear from a great height. Like the clove hitch, it increases the amount of friction on the rope but also allows it to slip, so it can be used instead of a belay.

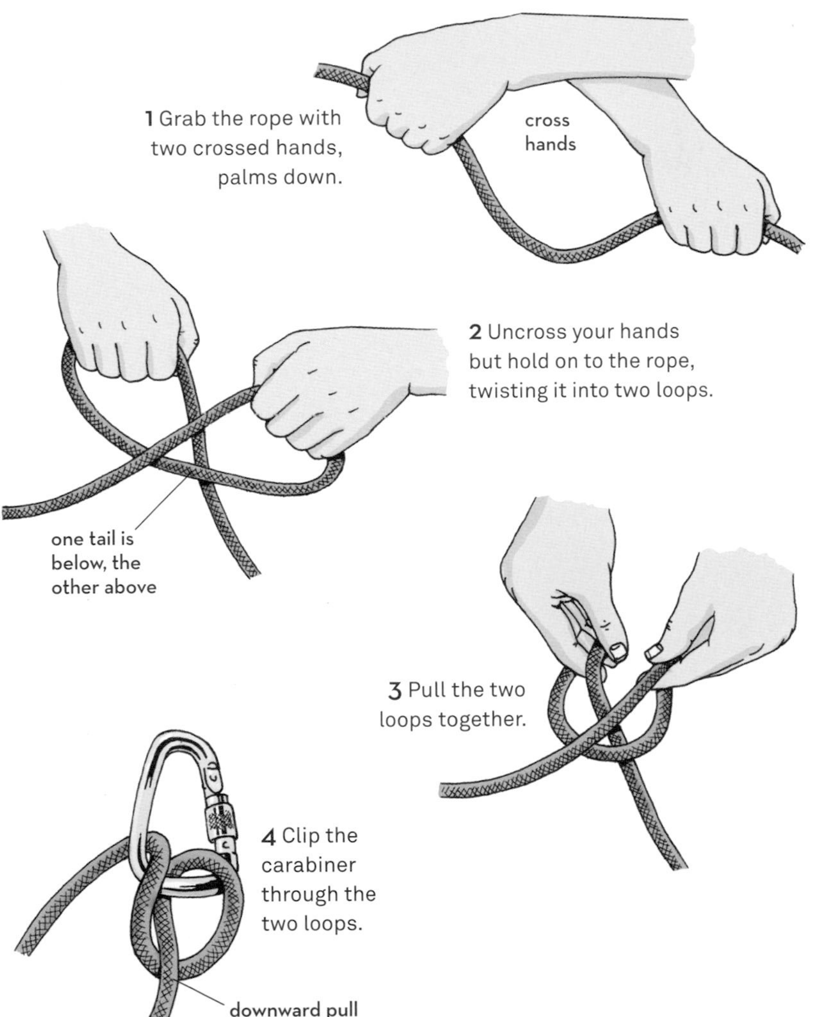

PILE HITCH

Pile hitches can be used to attach a dinghy to a post, or an anchor line to a bollard. They can be tied in the bight or at the end of a rope. The length of the working end required to tie the hitch will depend on the thickness of the post. To untie, ease some of the working end back into the hitch until there is enough slack to lift the bight off the top of the post.

1 Form a bight long enough to pass several times around the post. Hold the standing parts in one hand. With the other, wrap the bight around the post, underneath both standing parts.

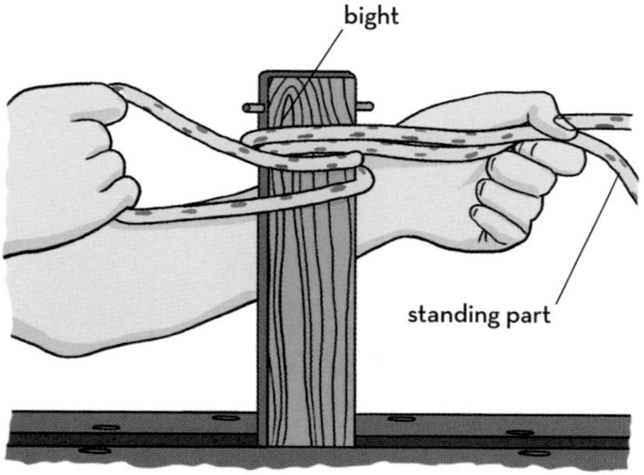

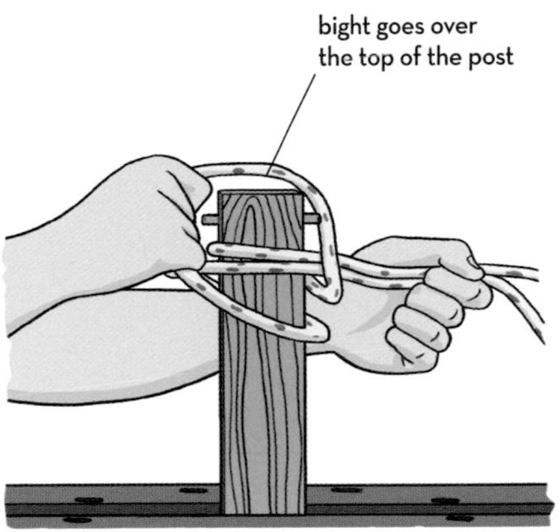

2 Widen the bight and place it over the top of the post.

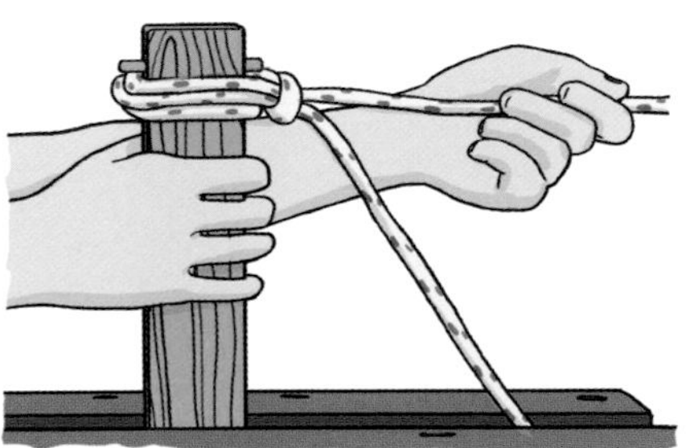

3 Pull evenly on both ends to firm the hitch. If the rope is sticking and not tightening evenly around the post, you may have to help it, but be careful not to catch your fingers.

Double pile hitch

The double pile hitch is very secure but a little more complicated to untie.

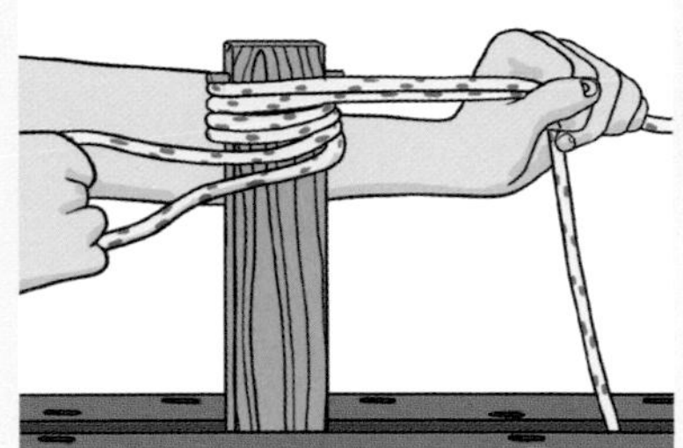

1 Complete step one of the pile hitch (shown left), then wind the bight around the post a second time, before placing it over the post. The second winding is made lower down the post than the first.

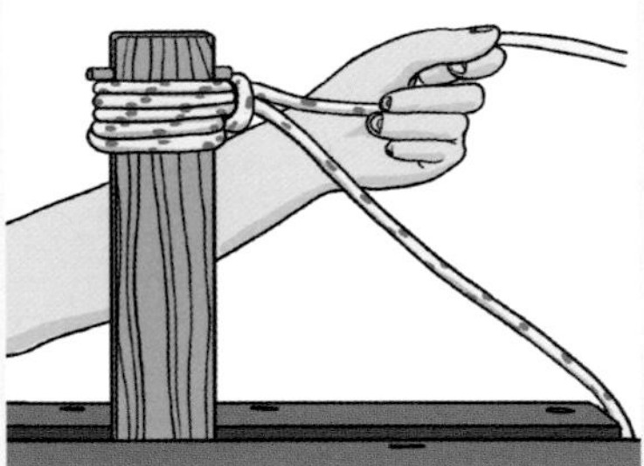

2 Adjust the hitch as in step three of the pile hitch. With both hitches, make sure that you hold the ends together until the bight is placed over the post, so that both ends are secure.

HIGHWAYMAN'S HITCH

The highwayman's hitch is a useful knot to know because it instantly releases with a tug on the working end. However, it is not the most secure of hitches and can be loosened quite easily.

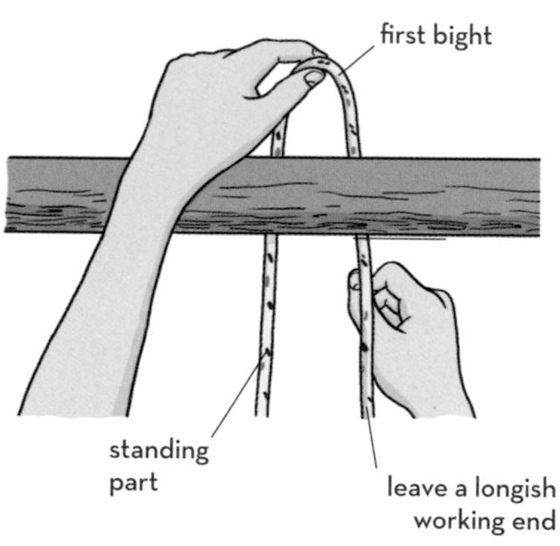

1 Form a long bight in the left hand, with the working end to the right of the standing part, and lead it up behind the rail.

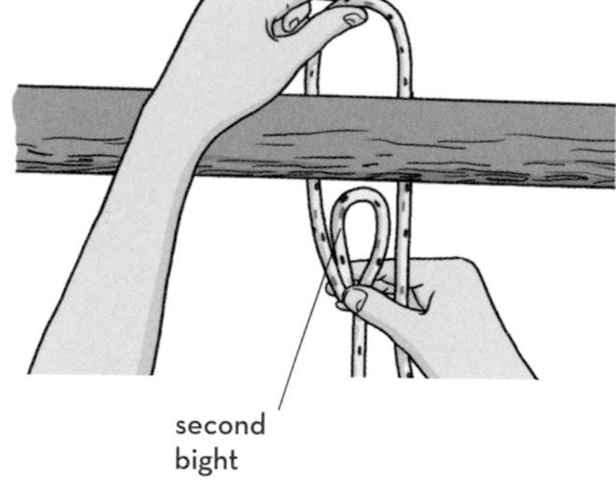

2 With the right hand, form a second bight in the standing part, with the remainder of the standing part on the right side of this new bight.

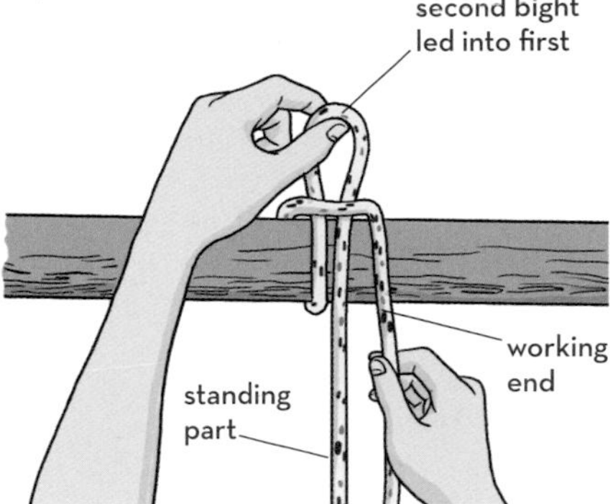

3 Tuck this second bight up through the first, hold it in place with your left hand, and tighten the first bight around the second by pulling on the working end.

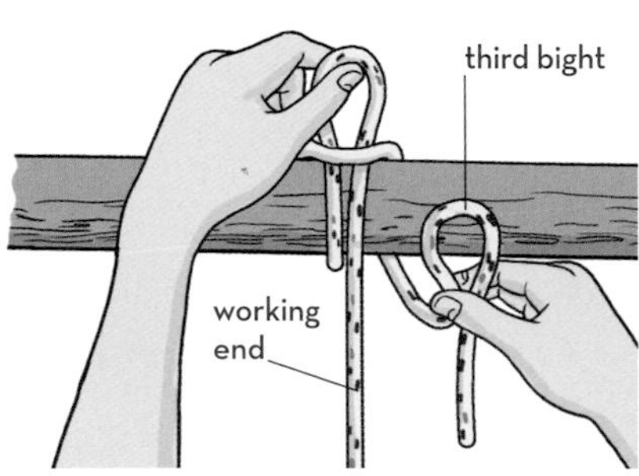

4 With the right hand, form a third bight in the working part, with the working end to the right.

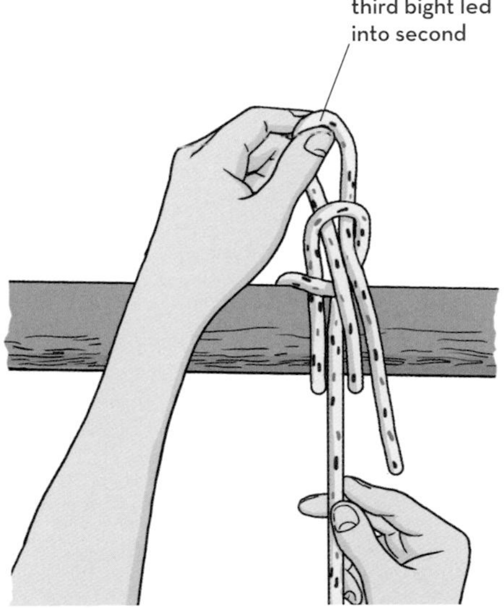

5 Lead this third bight fully up into the second, so that there is no slack around the rail.

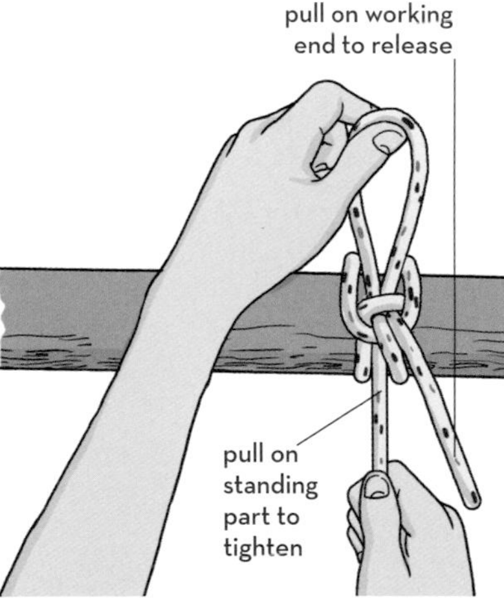

6 Pull on the standing part to firm the second bight around the third. To release, pull on the working end and the hitch will slide apart.

TIMBER HITCH

The timber hitch is an easy method of securing a rope to a pole. With the timber hitch tied at the center of gravity, the pole can be hoisted up. The addition of a half hitch forms the killick hitch, which allows the pole to be towed on land or through water. The timber hitch is also used to attach nylon strings to a guitar bridge.

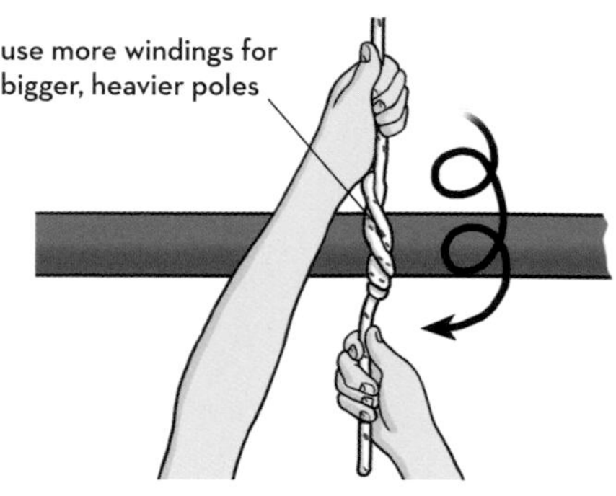

2 Wind the working end around the turn again once, twice, or three times. Pull tight on the standing part and ease the slack out of the windings through the working end.

1 Make a loose turn around the pole from the back to the front. Pass the working end behind the standing part, to the front, and under the turn.

Killick hitch

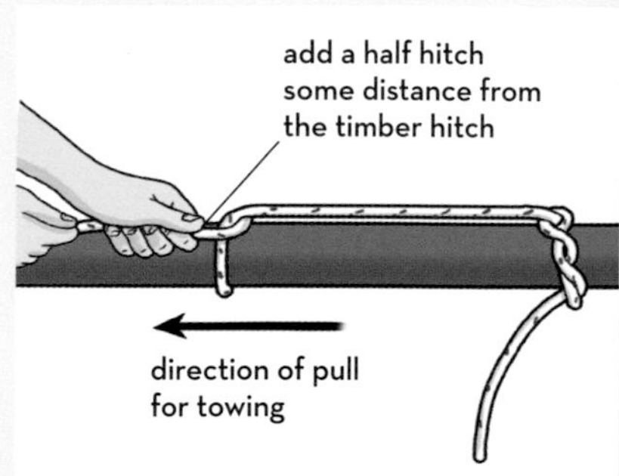

3 Lead the standing part along the pole and tie a half hitch. Tying it nearer the end provides better stability when towing, but too near the end and it will slip off.

ROLLING HITCH

The rolling hitch is used to tie a rope to a pole or to larger rope, when the load is to be applied at an angle between 45 and 90 degrees to the pole. The direction of the load, or strain, will decide the way in which the knot must be tied. The hitches shown here take the strain from the right.

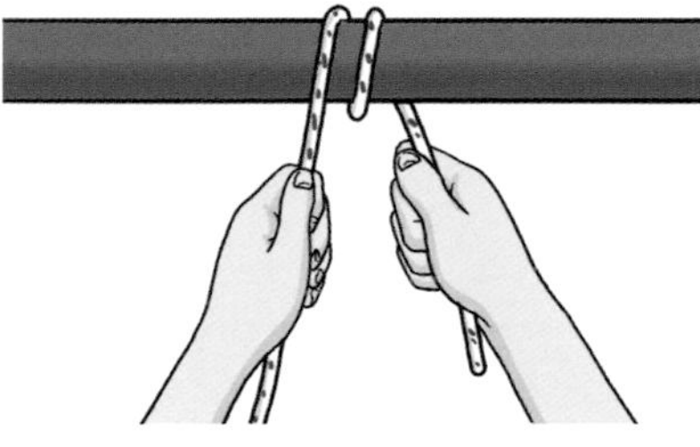

1 Begin the hitch with a round turn, going up and over, and from left to right.

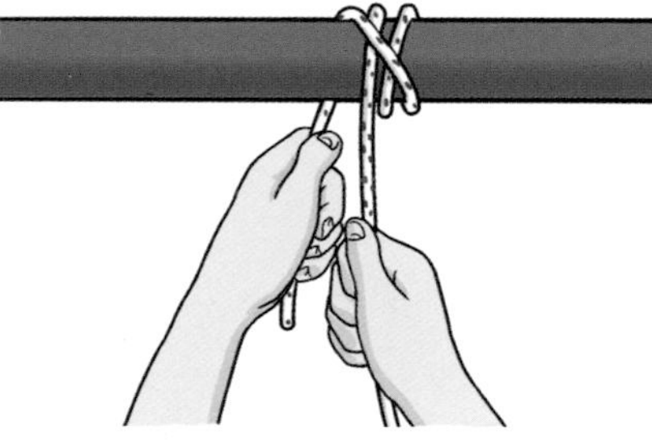

2 Pass the working end diagonally left across both windings, then down around the back of the pole.

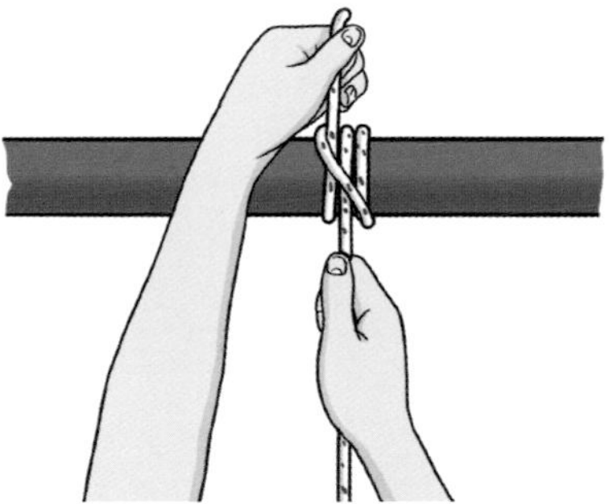

3 Tuck the working end up under the diagonal. Note that the rolling hitch is actually a clove hitch with an extra turn around the pole on the right.

BEAR SAYS

The tautline hitch is similar to the rolling hitch, and is suitable for attaching a line to a taut rope.

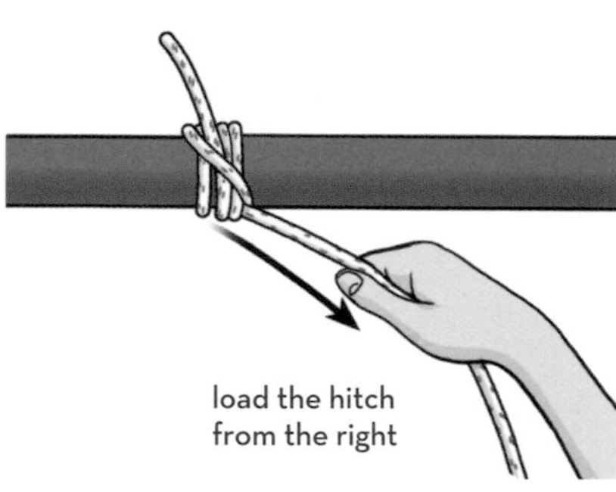

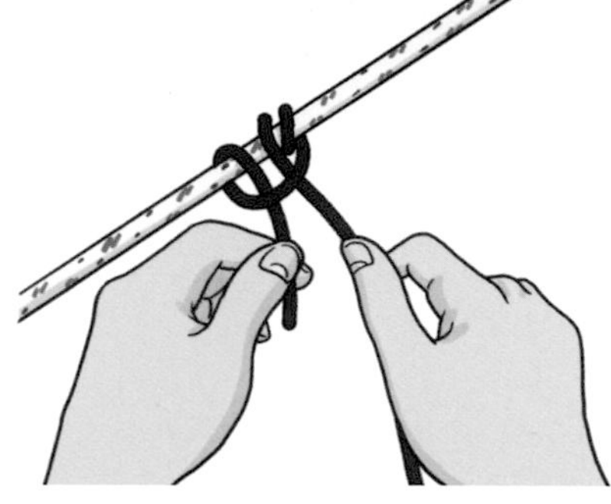

4 Pull both ends to tighten. The load can be applied from the right of the hitch. For loading from the left, begin tying the hitch as in step one, but winding from right to left.

5 Lead the working end up behind the rope, to the front, and tuck it under itself, parallel to the standing part. The knot looks like a cow hitch with an extra turn.

Tautline hitch

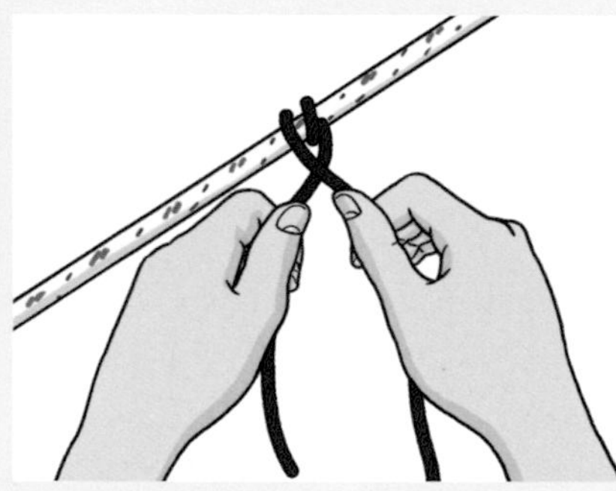

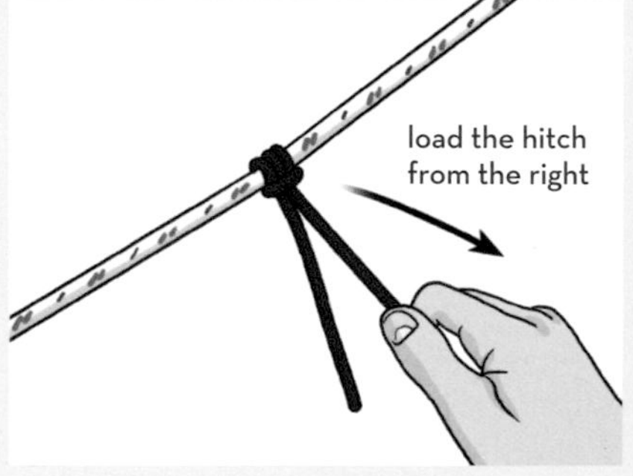

1 Tie this hitch using a line that is at least half the diameter of the taut rope. Complete step one of the rolling hitch, then lead the working end left in front of the standing part.

2 Pull both ends to firm the hitch tight around the rope. As with the rolling hitch, the strain can be applied from the direction in which the initial overhand turn was made.

TRUCKER'S HITCH

As it can be tensioned further after tightening, this knot is suitable for tying tent stays, or securing a load on a trailer. The hitch provides leverage, allowing the rope to be pulled tight, and uses up any excess cordage.

BEAR SAYS

The trucker's hitch is also known as the waggoner's hitch or dolly knot.

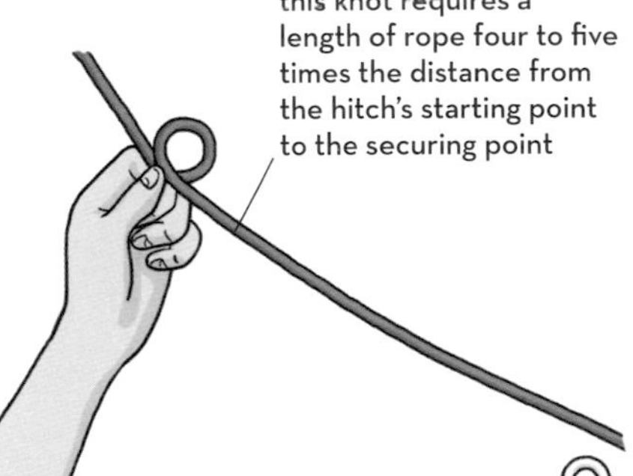

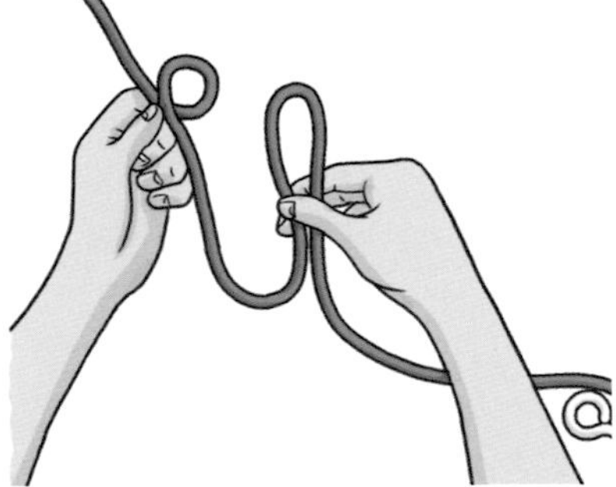

1 Begin with the standing part tied to an upper fastening point. Rotating counterclockwise, make a small overhand turn and secure it in your left hand.

2 With your right hand, form a bight in the working part. It will need to have a length approximately half the distance from the turn to the securing point.

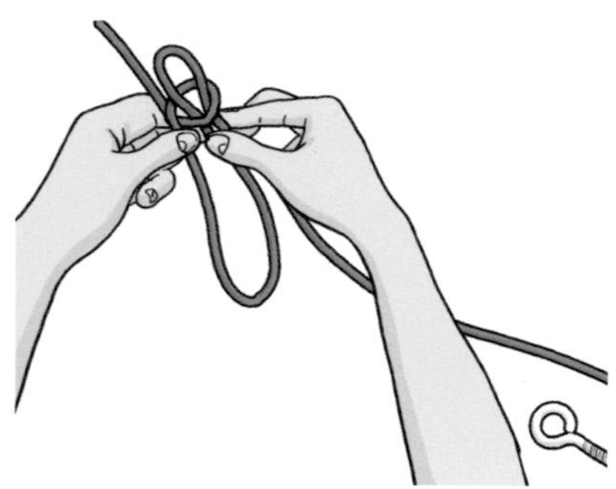

3 Lead the end of the bight into the overhand turn from below—not too far in, about a fifth of its length will do. Secure the bight and overhand turn in your left hand.

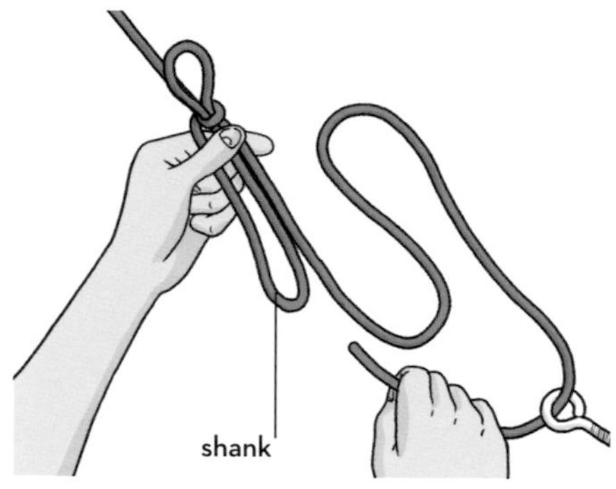

4 This forms a new, lower bight, called a shank. Lead the working end down and through the lower fastening point.

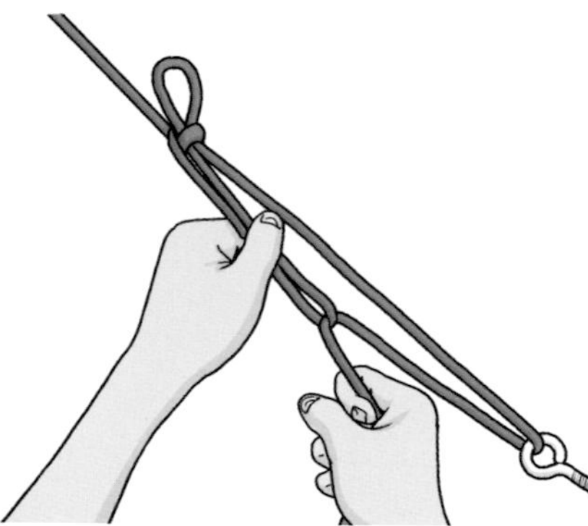

5 Lead the working end up through the shank from back to front. Apply downward tension to the working end. The turn will grip the bight and you can let go with the left hand.

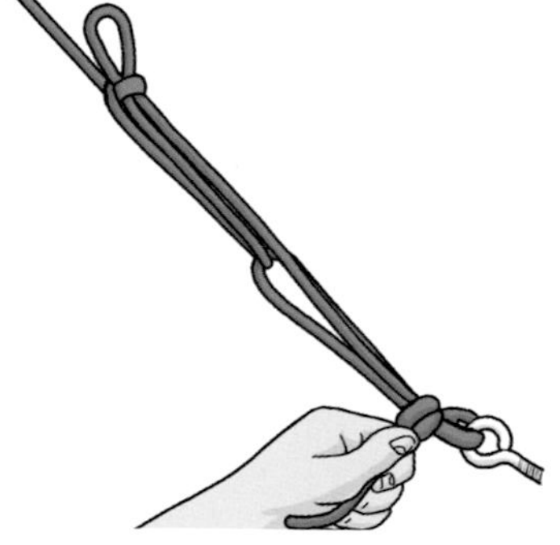

6 Pull the end as tight as necessary and tie it off above the fastening point—a couple of half hitches should do. The trucker's hitch requires some practice to master.

SQUARE LASHING

Square lashing is a relatively easy way to secure two poles at right angles. Be careful with the tension and the number of windings—the lashing must be strong enough for the job, but not so tight that the poles are bent.

Lashing is a method of fastening items together with cord. They are permanent and use multiple windings.

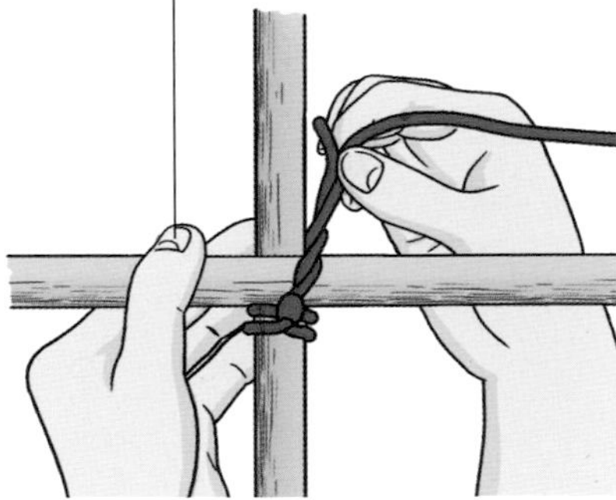

later additional windings will lock the tail securely

1 Tie a clove hitch to the vertical pole. Wind the tail and standing part together. Place the horizontal pole on top of the vertical. Lead the ends over both poles to the right.

BEAR SAYS

To join poles together, lashings are used. They can be useful around campsites to help build chairs and tables.

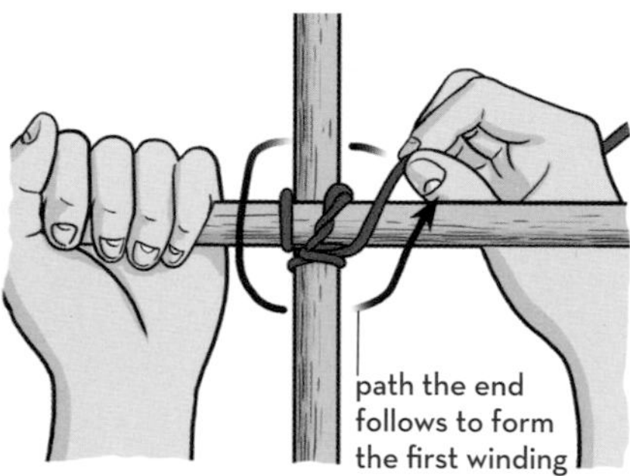

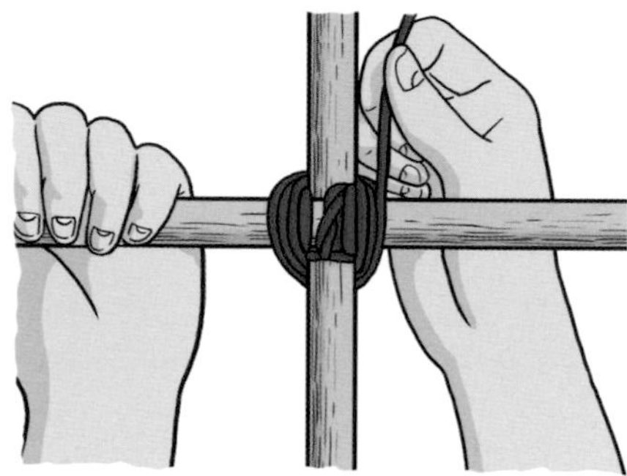

2 Maintaining the tension, lead the cord behind the upper vertical pole, over and down in front of the left horizontal pole, around behind the lower vertical pole and to the front.

3 Repeat the winding process about four times. The number of windings will depend on the diameter of the poles and the thickness of the cord.

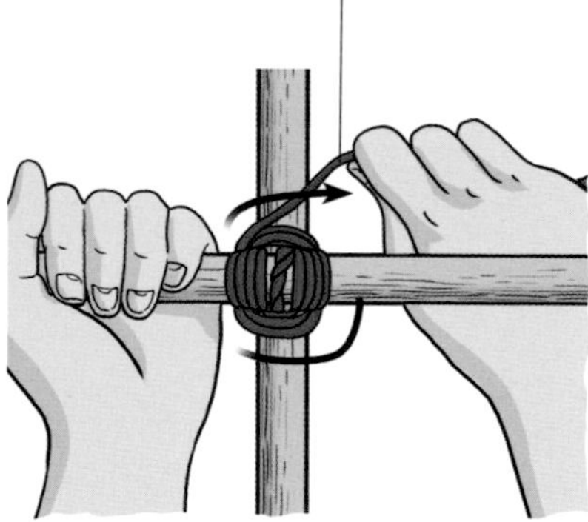

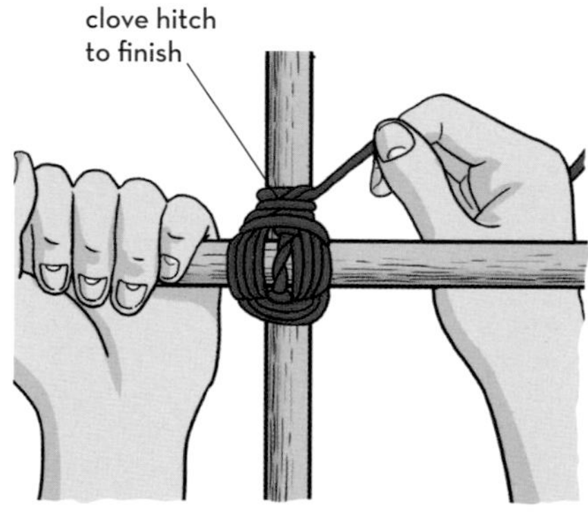

4 Start the frapping turns. Make a turn over the right horizontal pole, then wind clockwise between the two poles three or four times.

5 Stop at the top left and tie a tight clove hitch around the upper vertical pole, so the lashing can't slide or rotate under tension.

DIAGONAL LASHING

Diagonal lashing is used to secure poles that cross diagonally together. The two poles don't have to be at right angles—the timber hitch at the start pulls both poles together without changing their position. However, if the poles are not held in place, the angle can be difficult to keep during lashing.

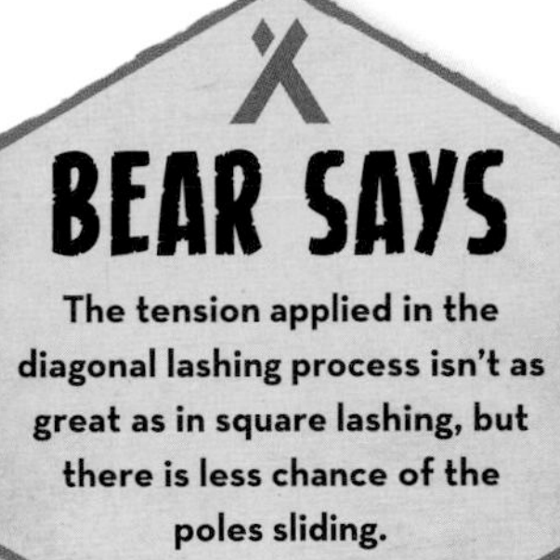

BEAR SAYS

The tension applied in the diagonal lashing process isn't as great as in square lashing, but there is less chance of the poles sliding.

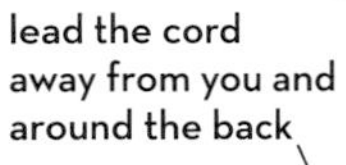

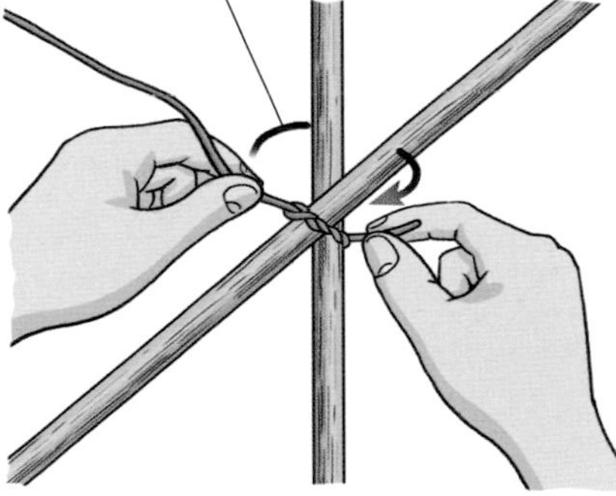

1 Tie a timber hitch around both poles, at the intersection with the widest angle. Tighten the hitch and lead the cord away from you, around the back of both poles.

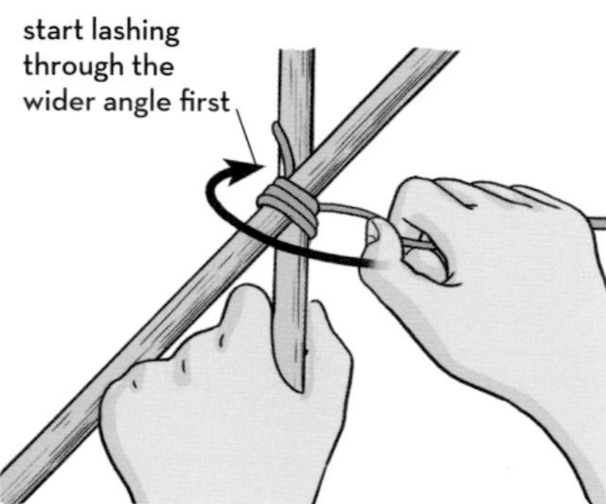

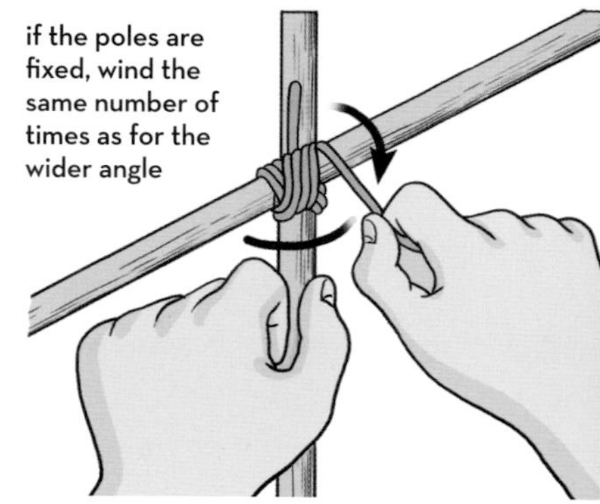

2 Wind tightly over the top of the hitch and around the middle four or five times. Unless the poles are fixed, the tighter you wind, the wider the angle becomes.

3 Now wind across the other, narrower angle. If the poles are not fixed, apply pressure and continue winding until you achieve the angle you want.

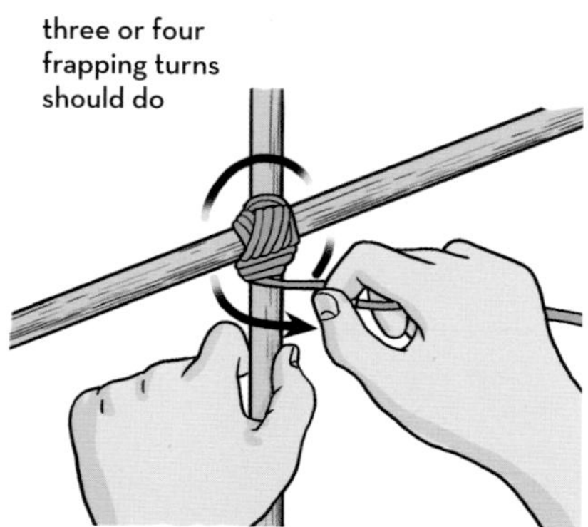

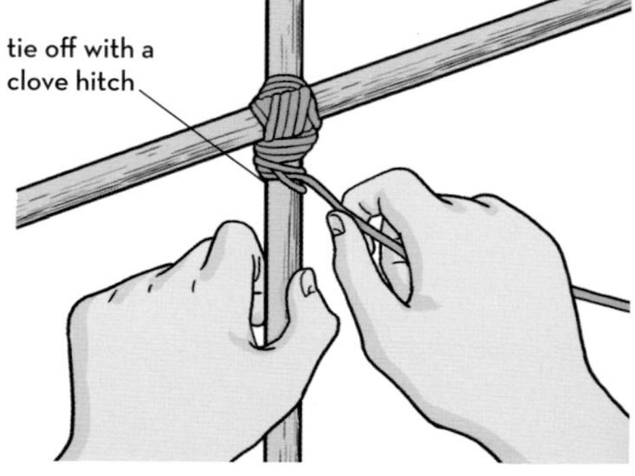

4 Wind the cord counterclockwise by passing in front of the upper vertical pole, behind the left cross pole, in front of the lower vertical pole and behind the right cross pole.

5 Finish with a clove hitch around one of the poles. Line it up with the end of the frapping turns so that there is little chance of the hitch sliding or rotating under tension.

SHEAR LASHING

A shear lashing can secure two poles together to reinforce them. It can also extend a pole if the lashing is placed near the end, at the point where the two poles overlap. Tied a little more loosely, it makes an A-frame lashing. Here, the two poles are separated slightly so that they can be moved apart. The A-frame is also known as shear legs and can be used for a tent or lean-to shelter.

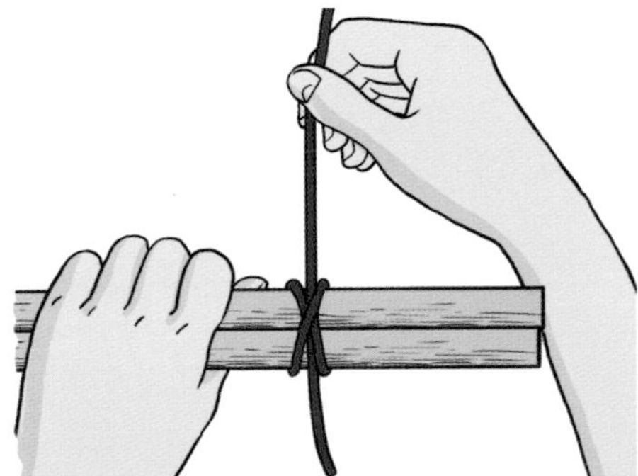

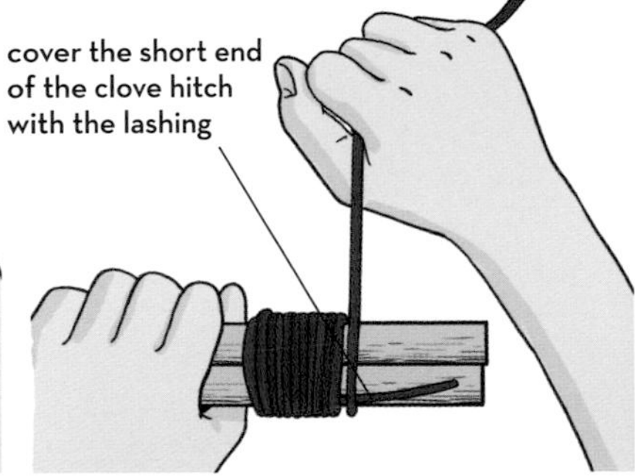

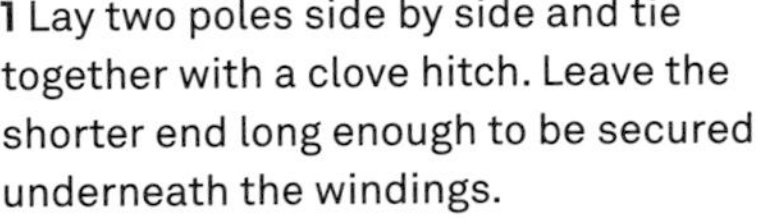

1 Lay two poles side by side and tie together with a clove hitch. Leave the shorter end long enough to be secured underneath the windings.

2 Begin winding, but not as tightly for an A-frame as for securing two poles. As a rule, make the binding length no less than the width of the two poles.

3 For the frapping turns, lead the cord behind the top pole and to the front between the poles. Wind across the existing windings, between the poles.

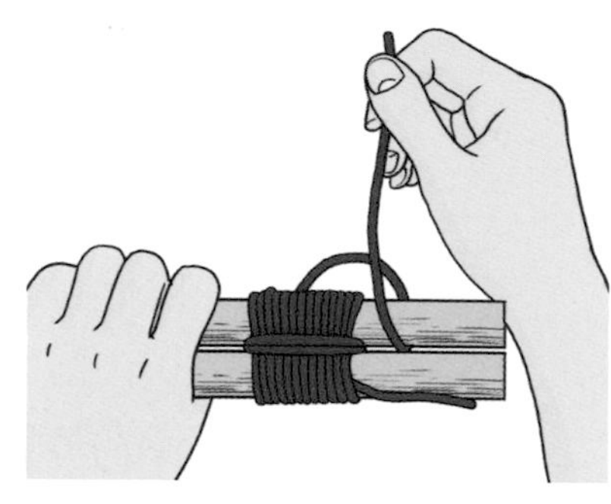

4 Finish at the opposite end to the original clove hitch and tie another clove hitch around one pole, not both. The hitch must be tight, and snug against the lashing.

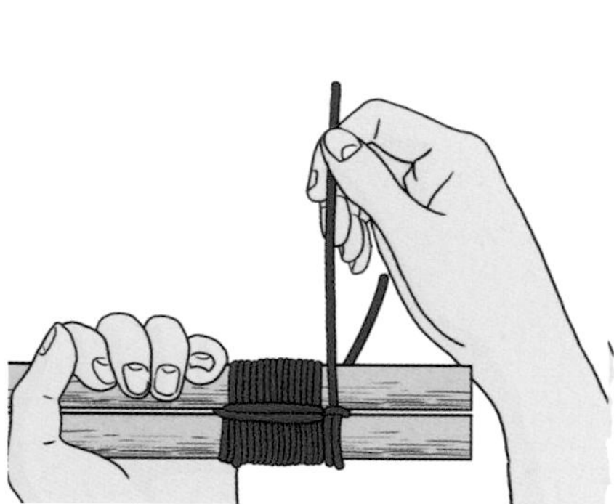

5 If tying poles together to reinforce or extend them, make a tighter binding, leave out the frapping turns, and finish with a clove hitch around both poles.

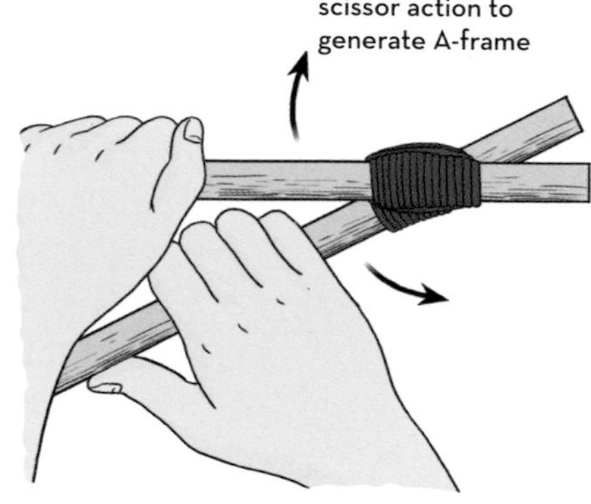

6 To use as an A-frame, separate the poles by using a scissor action, stretching the cord equally at both ends of the lashing. Practice will help you apply the right lashing tension.

TRIPOD LASHING

There are different methods of tying tripod lashings but this one can be tied into a frame, taken to a site, and put up. After use it can be folded flat and taken away, with a temporary binding (such as the pole lashing) securing the other end. However, it doesn't form the perfect triangle shape at its base. Like the shear lashing, the angle at which the legs can be separated depends on the length and tension of the lashing, and the stretch in the cord.

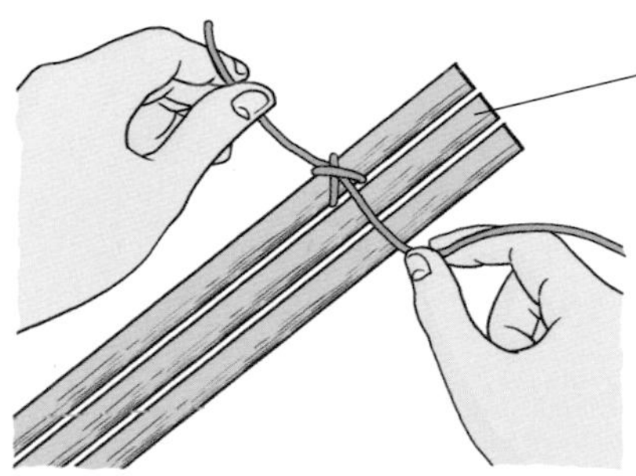

when beginning, leave space for the width of the lashing plus enough space to hang an item

1 Lay three poles side by side, making sure the ends that will stand on the ground are even. Tie a clove hitch around the top pole at a suitable distance from the ends.

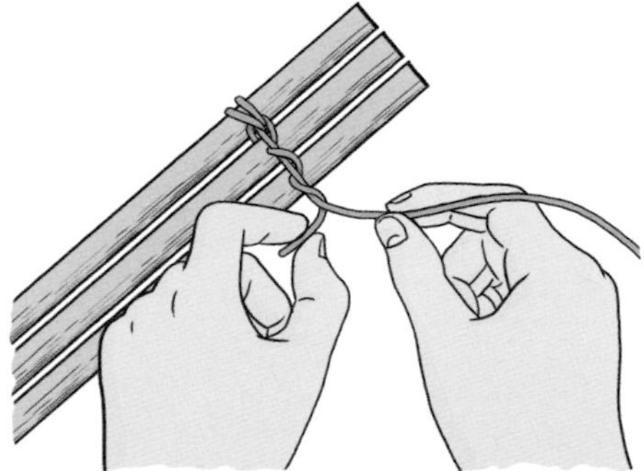

2 Wind the short clove hitch end around the working end and lead both ends toward you across the three poles. This will help lock the clove hitch and its tail.

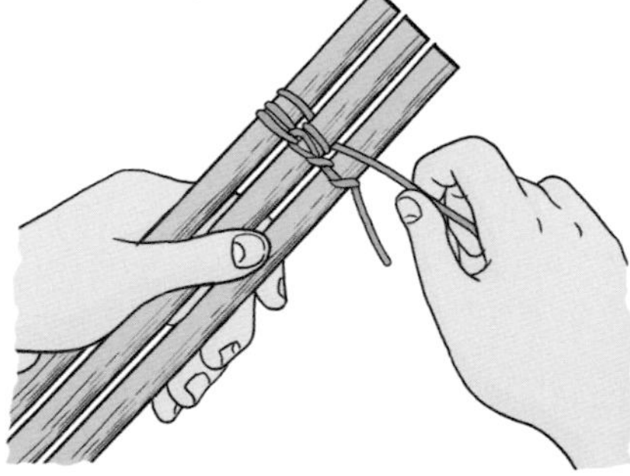

3 Start winding between the poles: under the bottom pole, over the center, under the top, then around and over, this time under the center pole, and over the bottom.

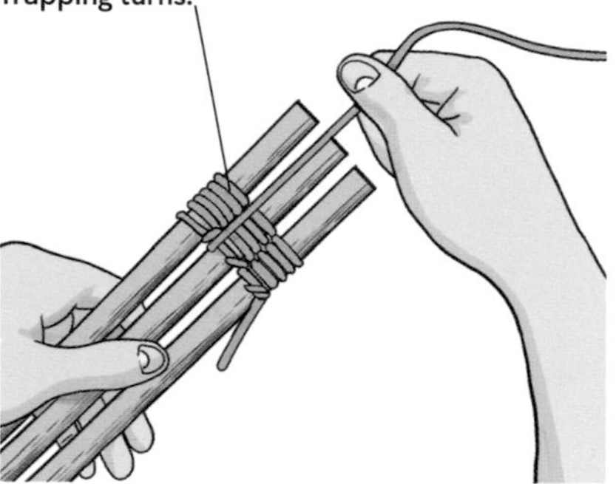

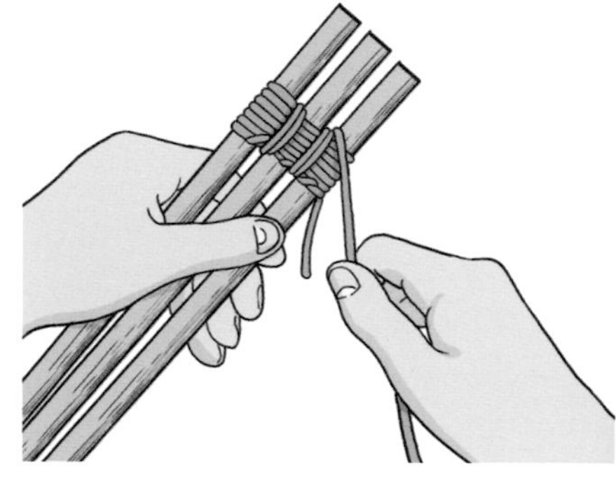

4 Begin frapping turns by leading the cord down in front of the top pole, then to the back between the top and center poles. Form two or three turns around the windings.

5 Lead the cord behind the center pole, and to the front between the center and bottom poles. Make a second set of frapping turns in the opposite direction to the first.

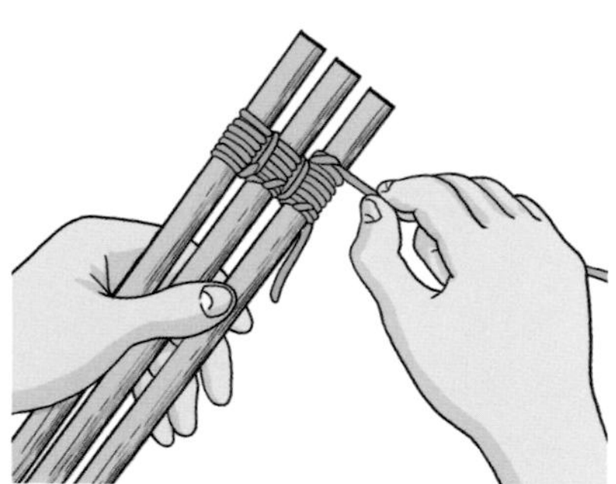

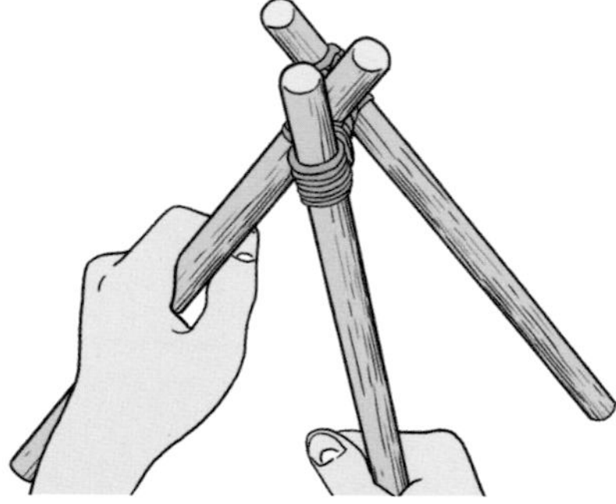

6 To finish, tie off with a clove hitch around the bottom pole. The frapping turn must lead straight into the hitch so that there is no chance of it rotating.

7 To erect the tripod, separate the outer poles and use a scissor action to swing the center pole in the opposite direction. This may be difficult if the lashing is too tight.

EYE SPLICE

Rope splicing is a way of joining two pieces of rope by unraveling their strands and then weaving them together. A splice is used where a rope is fixed permanently to an item or slipped over a hook, stake, or bollard. An eye splice isn't actually a knot, but a way to form an "eye" in the end of a length of laid rope.

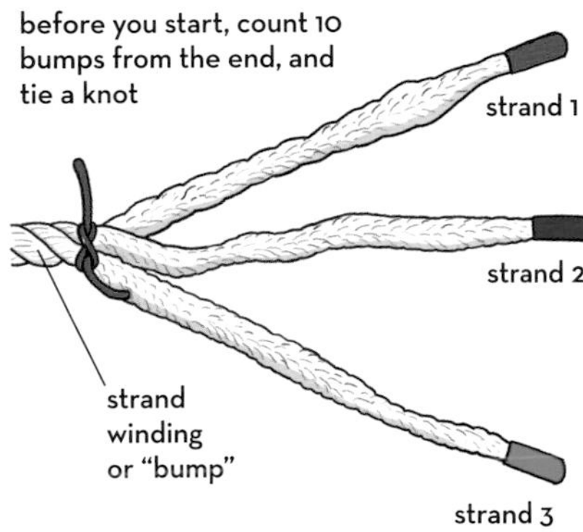

Splicing is a traditional method used to permanently join two laid ropes.

1 Begin to unlay the rope. As you separate the strands, bind them with tape. Continue to unlay the strands up to the knot. Number each strand.

2 Open the lay of the rope to raise one strand. Feed the central strand, strand 2, diagonally left under the raised strand in the standing part. Do not pull it through completely yet.

3 Feed strand 3, which is to the right of strand 2, under its corresponding right strand in the standing part, that is, the strand to the right of the one that strand 2 is tucked under.

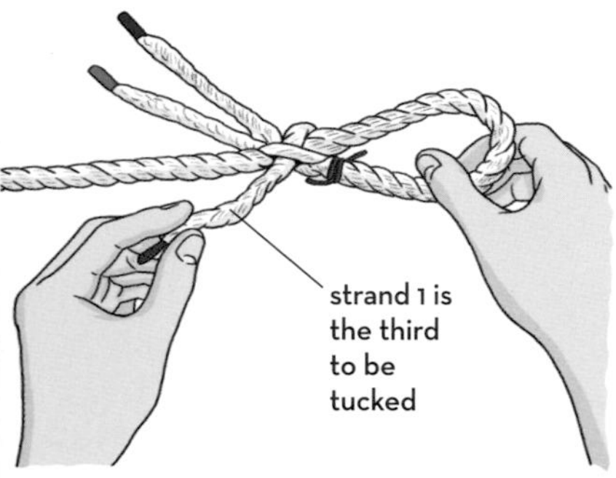

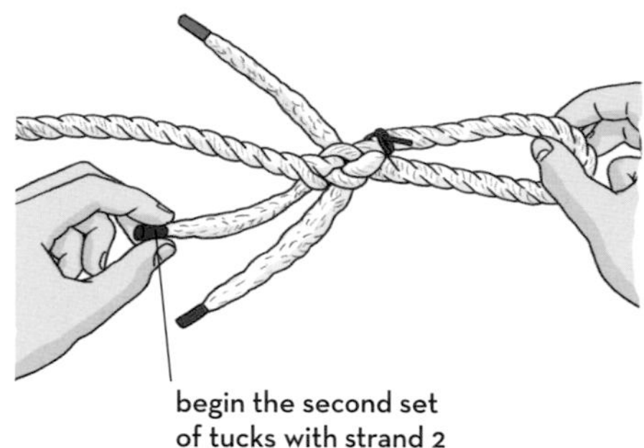

4 Rotate the rope slightly, and tuck the remaining strand 1 under its corresponding left strand of the standing part. Pull the three unlaid strands snugly up to the standing part.

5 Continue weaving the unlaid strands in this diagonal pattern until you have completed four sets of tucks. Three may be enough, but not if the lay of the rope is loose.

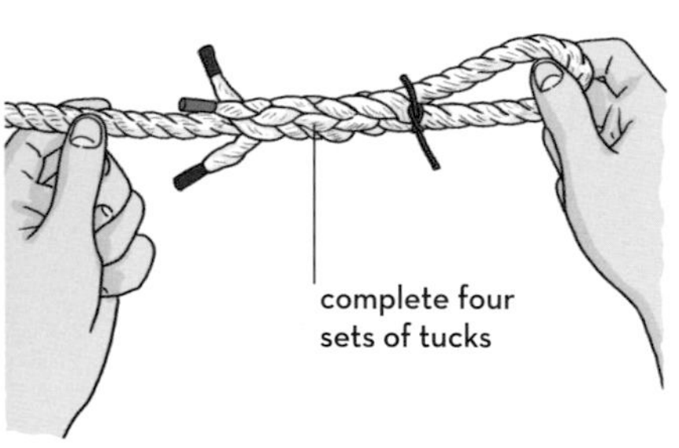

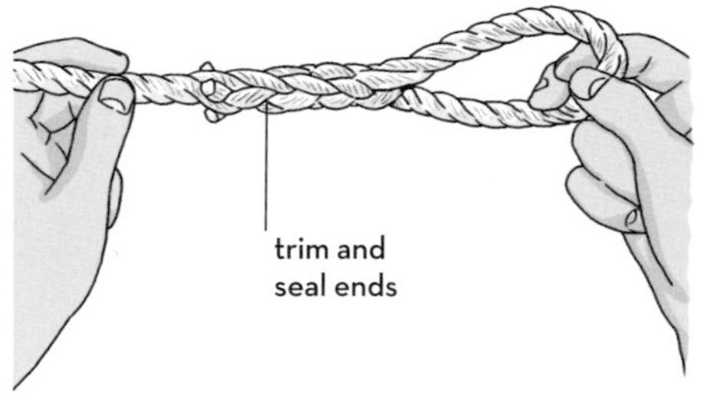

6 Maintain the even twist in the unlaid strands as you tuck. If you don't open the lay in the standing part sufficiently, the twist in the strands will increase as you go.

7 Thick rope may need the assistance of a fid (a cone-shaped tool) for opening the lay. After the fourth tuck, the ends can be trimmed and either sealed with heat or whipped (bound with twine).

SHORT SPLICE

This is a method of permanently joining two lengths of laid rope of a similar thickness. This method avoids the rope becoming thicker at the join.

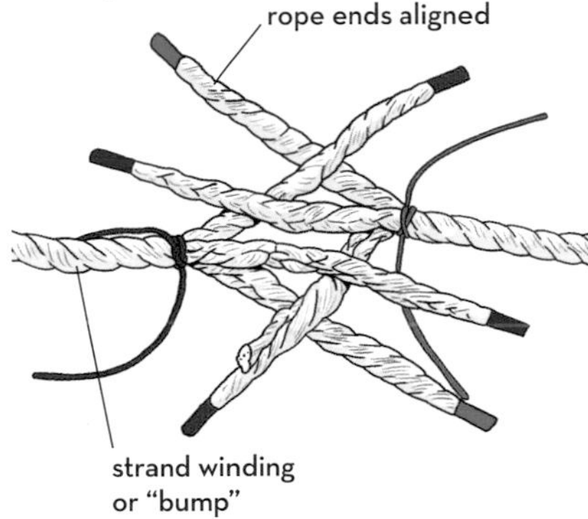

BEAR SAYS

Although it looks complicated, if you color code the strands, the technique becomes far simpler.

1 Tie a constrictor knot around each piece of rope about 12 bumps from the end. Tape and number the strands and unwind them. Align the strands of the rope ends.

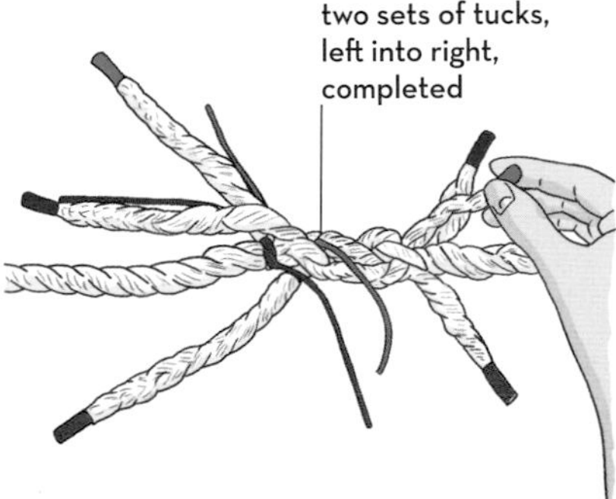

2 Following steps two to four of the eye splice (p.135–36), begin to weave the left rope into the right. Make two sets of tucks.

3 Make a set of tucks of the right rope into the left. Tighten the strands and adjust the alignment of one rope to the other. Loosen the constrictor knots as needed.

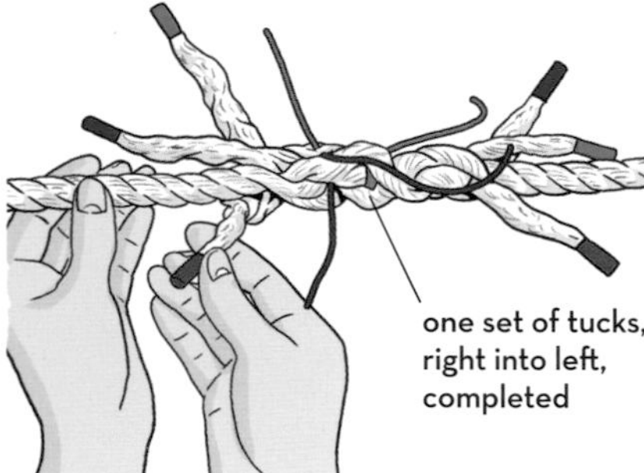

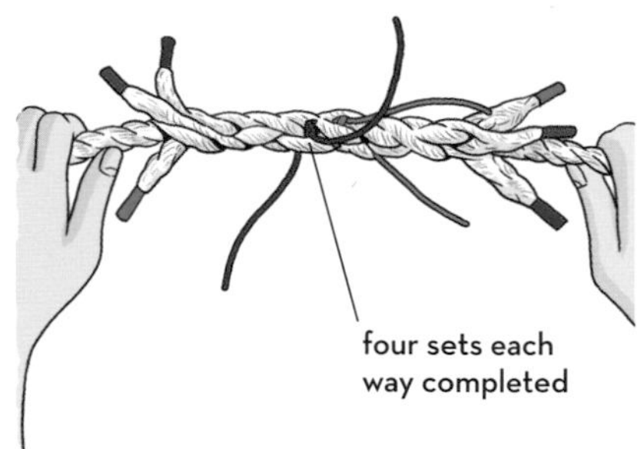

4 Make another set of tucks with the right rope into the left, so that there are two sets in each direction. Then make alternate sets until you have four in each direction.

5 The splice is completed and the ends can be trimmed and sealed. However, if you wish to taper the splice so that it blends neatly into the rest of the rope, proceed to step six.

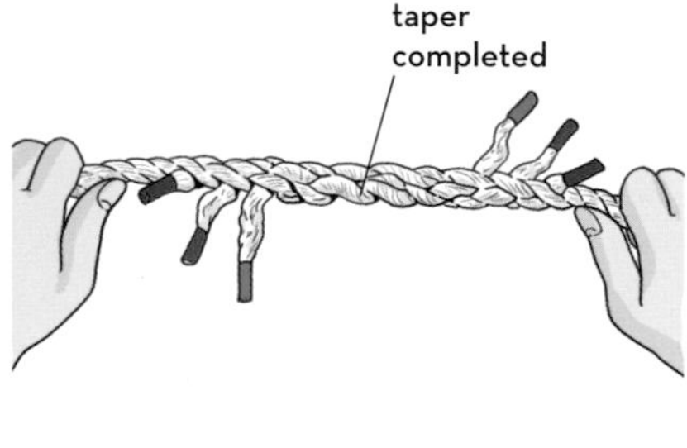

6 An easy way to taper is to not tuck one strand of each set after three tucks, leave out a second strand after the fourth tuck, and make a fifth tuck with the one remaining strand.

7 When the tucking of each strand is completed in the correct order, the splice will have an even taper. Trim and seal the ends neatly.

IMPOSSIBLE KNOT

Put a length of rope in front of your friends and challenge them to pick it up then tie a knot without letting go with either hand. The result must be a proper knot that does not collapse when the rope is pulled tight.

BEAR SAYS

The trick behind mastering the impossible knot is plenty of practice, so you can demonstrate it quickly!

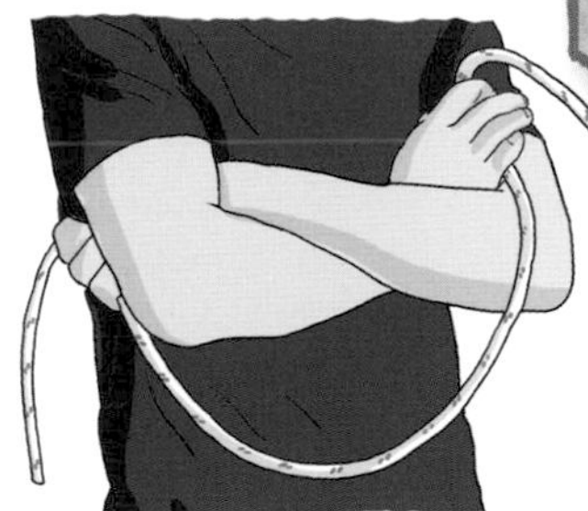

1 Lay the rope on a table. Before picking it up, cross your arms over each other. Now lean over and pick up one end of the rope in each hand.

Rope tricks are ways of tying and untying a knot that at first glance appear impossible!

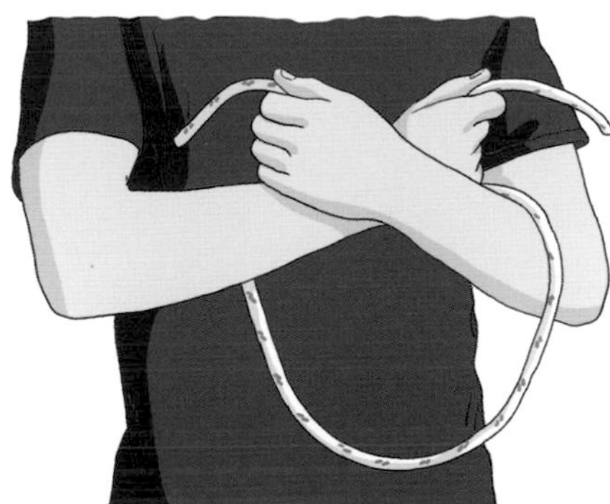

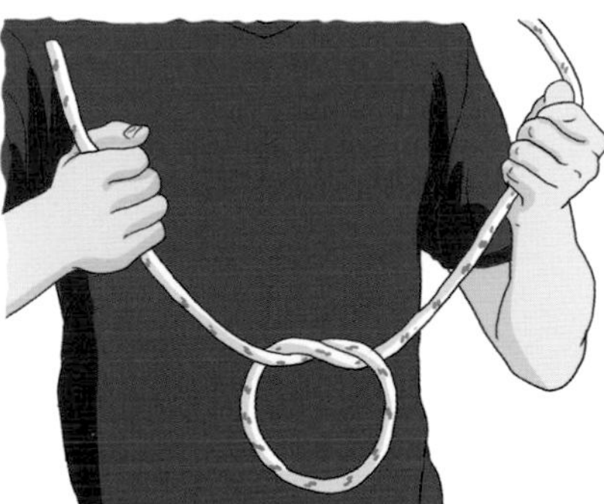

2 Keeping a firm hold on the rope, uncross your arms and move your hands apart, letting the rope slip over your wrists and hands as you do so.

3 With this trick, you tie an overhand knot in the rope without letting go of either end.

RING DROP

This trick allows you to remove a ring that has been threaded onto a loop. The loop should be made from a length of thin cord about 3 ft. long. Practice this first so you know how to do it quickly, then amaze your friends by getting them to try first—then if they can't figure it out, show them how it is done!

1 With the loop hanging over both your thumbs, hook the right little finger around the upper strand, to the left of the ring and from behind.

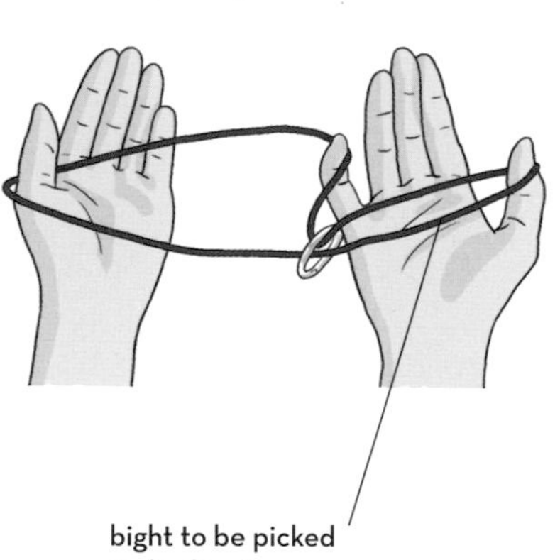

2 With your left little finger, reach over the bight formed by the right little finger, and hook it around the upper strand to the right of the ring, again from behind.

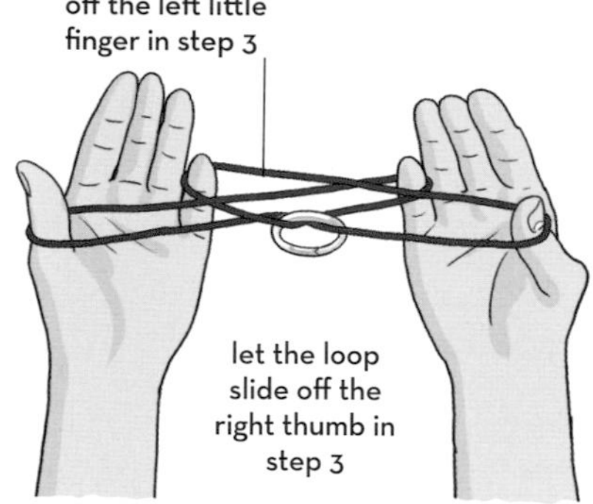

3 To free the ring, move your hands apart while letting the loop slip off the left little finger and the right thumb.

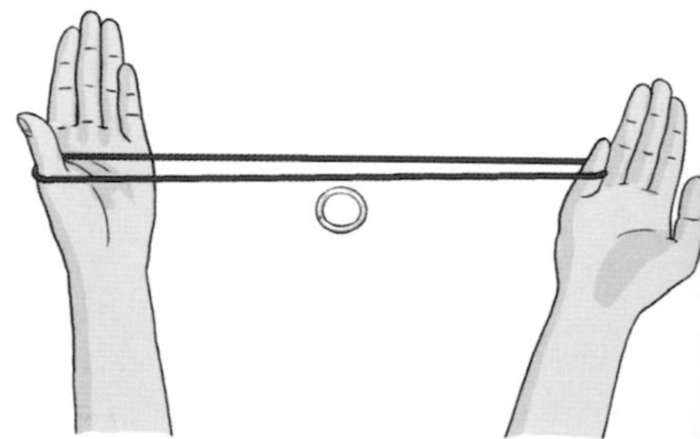

NOT A KNOT

This trick, which begins with a reef knot, looks very complicated, but the knot falls apart completely when pulled tight. The final step is easiest to carry out if thin cord or flexible rope is used rather than thick or stiff rope. A piece of cord about 3 ft. long is ideal.

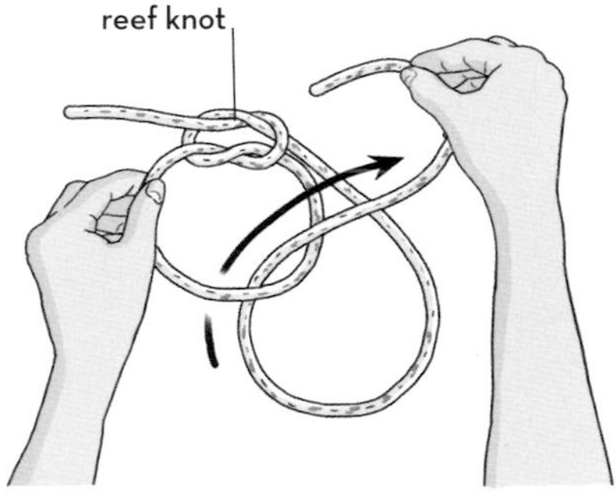

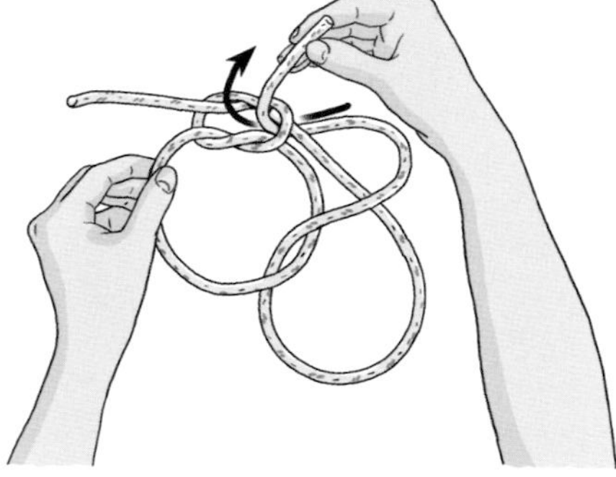

1 Form a loop and tie a reef knot. Pick up the working end that emerges toward the back of the knot and lead it through the loop from back to front.

2 Lead the same working end through the center of the reef knot, again from back to front.

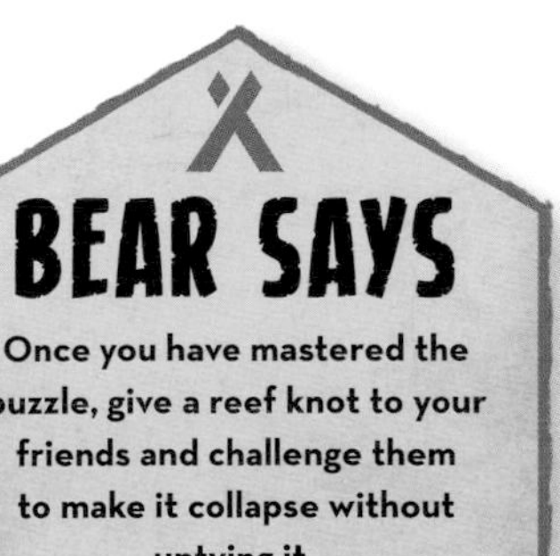

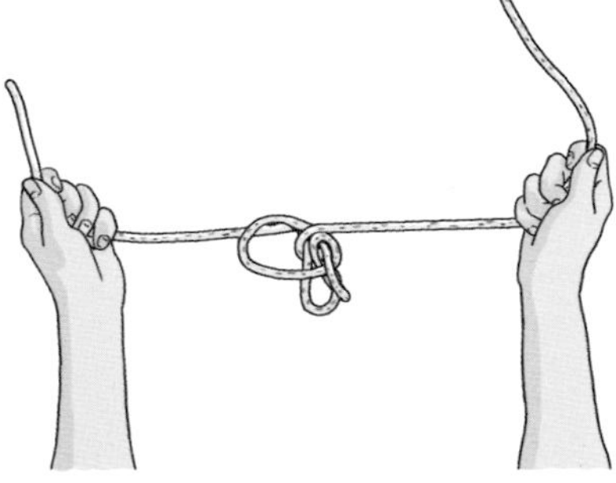

3 Pull firmly on both ends of the cord. The original reef knot will collapse and disappear completely.

DANGERS AND EMERGENCIES

There is so much out there in our amazing world to explore and experience, yet it is vital that you remain unharmed as you pursue new adventures. Our planet and its creatures can be dangerous at times, so learn the skills you need, and remain safe in the face of adversity.

Bear

IN THIS SECTION:

STAY SAFE IN THE WILD

When you set out on an adventure it is very important to be fully prepared in case you come into contact with danger. There are many ways to get help and avoid harm, so do your homework and stay safe.

Signaling for help

In an emergency your first contact with the outside world is likely to be a search aircraft. Make this contact count by learning standard ground-to-air signals. You can use objects, as well as your own body, to seek help.

All is well

Pick us up

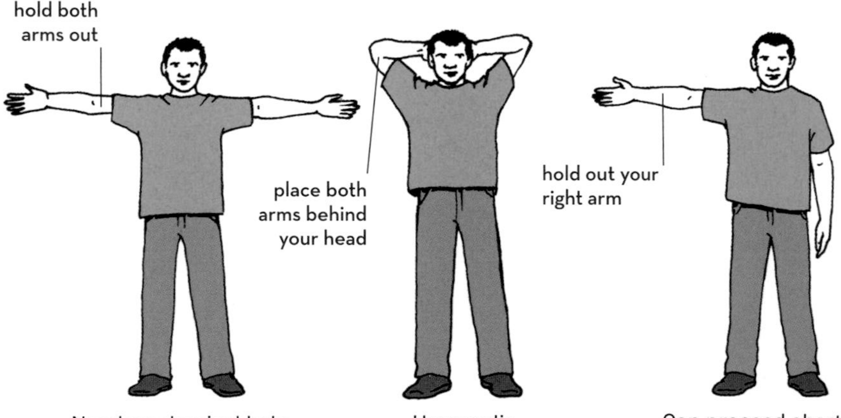

Need mechanical help

Have radio

Can proceed shortly

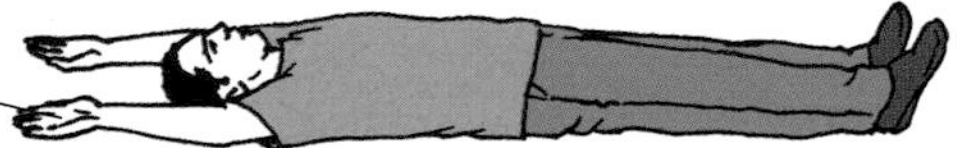

Need medical help

Do not attempt to land here

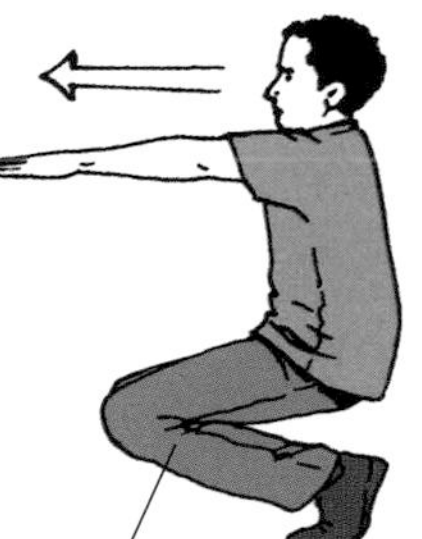

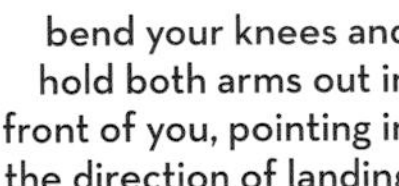

Land here

BEAR SAYS

If you are in distress, these signals could save your life. Choose a large, open area where you are most likely to be seen.

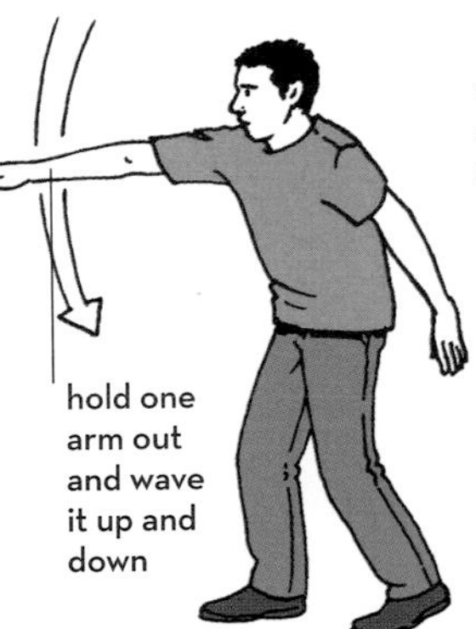

Use drop message

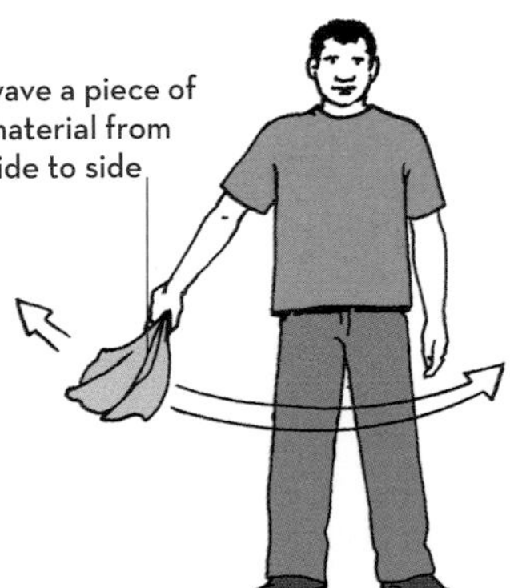

Negative (no)

Affirmative (yes)

Fire cones

Keep these primed with plenty of dry fuel, and ready to go at all times.

Smoke cones

In the daytime, smoke marks your position more clearly than fire. Use fuel sources such as green branches and rubber.

BEAR SAYS

Wherever you are in the world, three objects together signal distress. Don't forget this international call for help!

Life rafts

In thick jungle, the only clear area may be a river. Tether together three rafts loaded with fuel for a jungle distress signal.

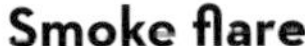

Smoke flare

flare smoke is usually red or orange and is designed to be seen up to 3 miles away
end cap with pull cord firing mechanism inside

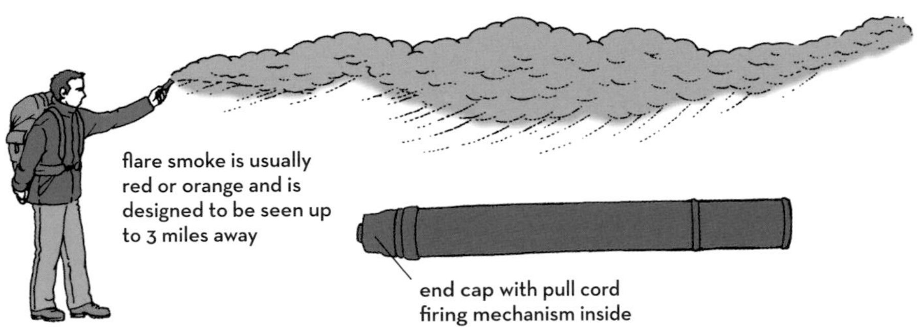

Other types of signals

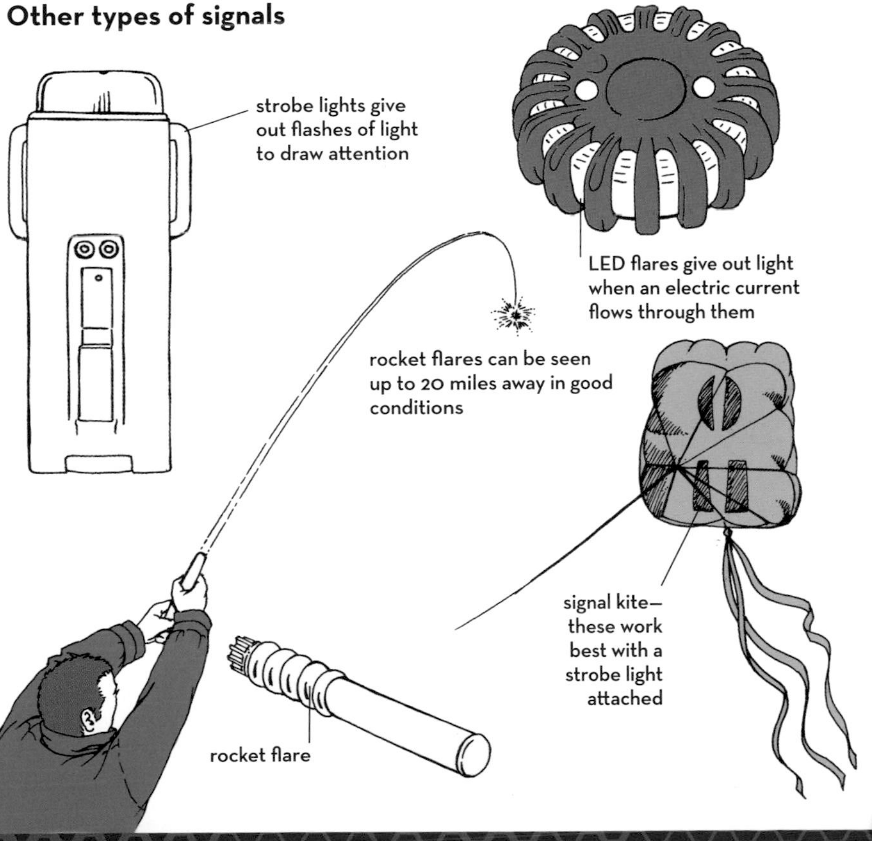

strobe lights give out flashes of light to draw attention
LED flares give out light when an electric current flows through them
rocket flares can be seen up to 20 miles away in good conditions
signal kite—these work best with a strobe light attached
rocket flare

Signal mirror

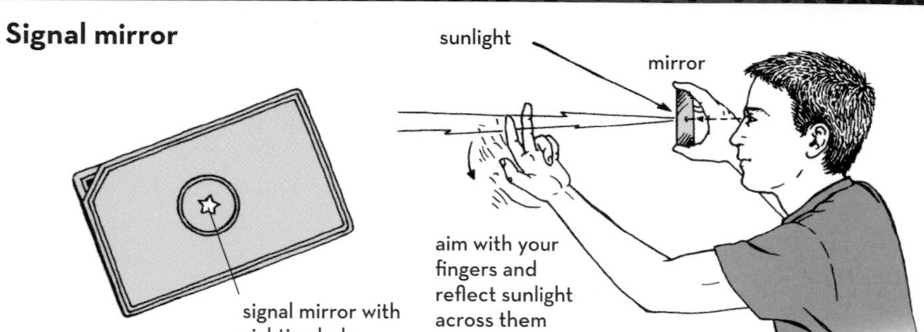

A mirror is the most valuable means of signaling in daylight, as reflections can easily attract attention. Make sure to pack one if you are going off the beaten track.

Personal locator beacons

High-tech personal locator beacons (PLBs) are small, lightweight devices that can be used in an emergency anywhere in the world. First, the PLB is activated (1), then a signal is sent to a satellite network in space (2). A ground station then receives the signal relayed by a satellite (3). The search and rescue coordination center is alerted (4) and then help is sent (5).

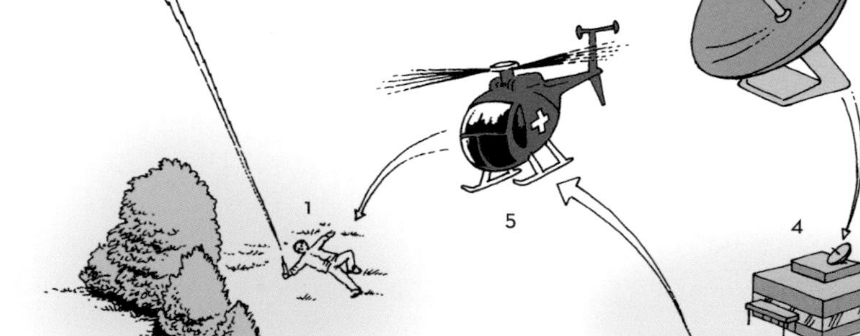

Aircraft signals

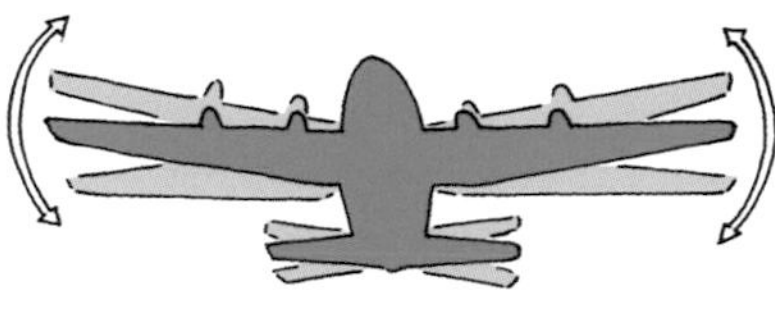

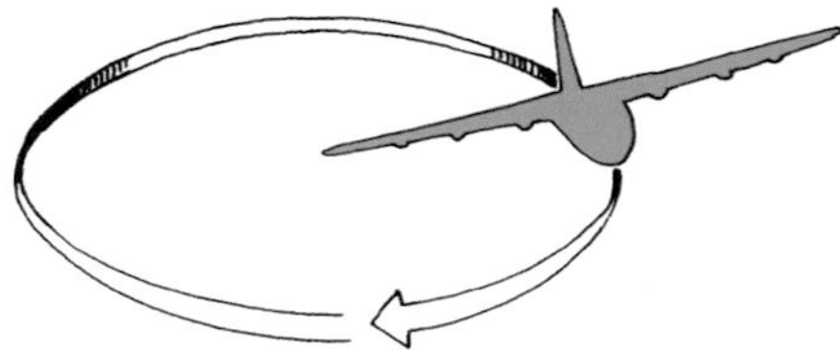

A rocking wing

If a plane rocks its wings the pilot has understood your signal.

Flying clockwise

This maneuver indicates that your signal is not understood.

Prepare a helicopter landing zone

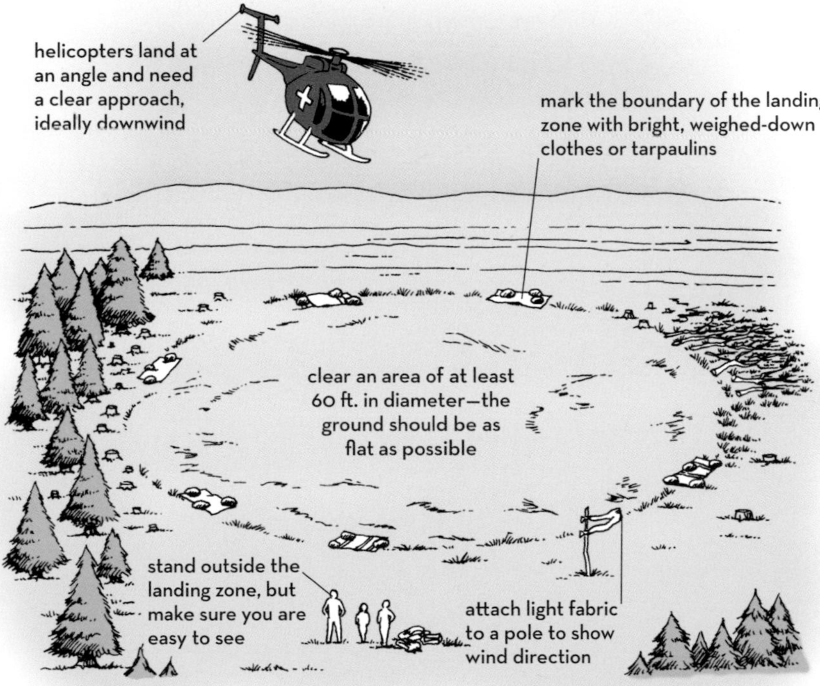

helicopters land at an angle and need a clear approach, ideally downwind

mark the boundary of the landing zone with bright, weighed-down clothes or tarpaulins

clear an area of at least 60 ft. in diameter—the ground should be as flat as possible

stand outside the landing zone, but make sure you are easy to see

attach light fabric to a pole to show wind direction

DANGEROUS ANIMALS

Most creatures will avoid human contact, but it pays to know what species are dangerous and what your defenses are.

Insects

Insects are small, often winged animals with six legs. Most are harmless, but some can be deadly.

Bee

A bee sting is painful but only life threatening to those who are allergic. If you are attacked by a swarm of bees, run away from the point of first contact, protect your face, and seek shelter.

Ant

Ant stings range from harmless to agonizing. Be sure to avoid the bullet ant of Central and South America. Its sting is considered the most painful of any bee, wasp, or ant.

Wasps and hornets

Relatives of bees and ants, these insects can sting over and over. They can be aggressive when seeking food, and are drawn to sweet odors. Stay away from nests as allergic reactions can be fatal.

Mosquito

The mosquito is one of the deadliest creatures on Earth. Mosquito-borne diseases are a big problem in the tropics, but can occur in temperate regions too.

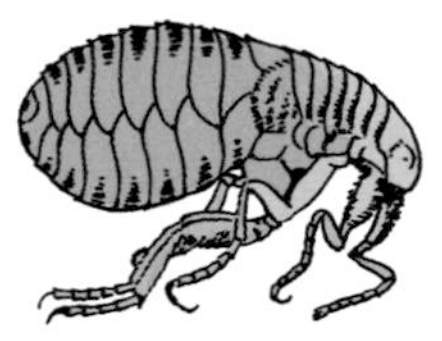

Flea

A flea bite is normally just an irritation, but they can lead to many diseases, including Lyme disease and even bubonic plague. It is sensible to consider them as a threat.

Tsetse flies

These large, bloodsucking insects are found in Africa between the Sahara and the Kalahari deserts. They carry the parasite that causes sleeping sickness, which can be fatal.

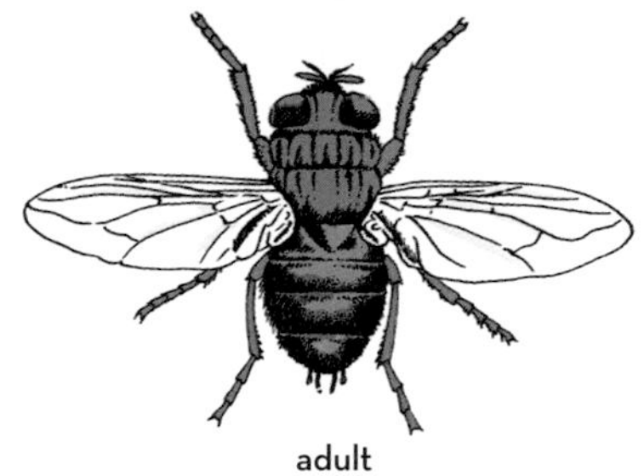
adult

larva

Botfly

An egg of the human botfly (native to Mexico, Central, and South America) hatches when it detects human warmth. The larva then burrows into the skin where it grows for about eight weeks. They may cause painful swellings but are otherwise harmless.

How to remove a botfly larva

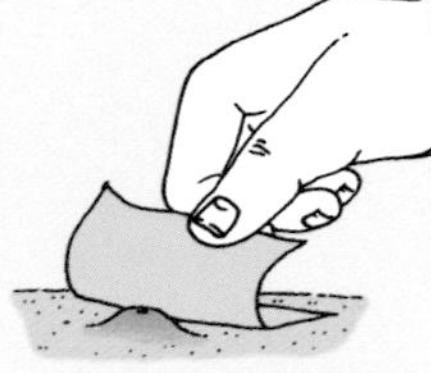

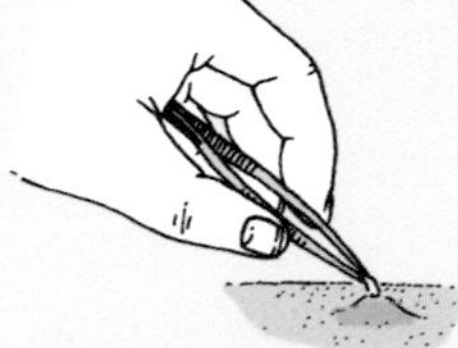

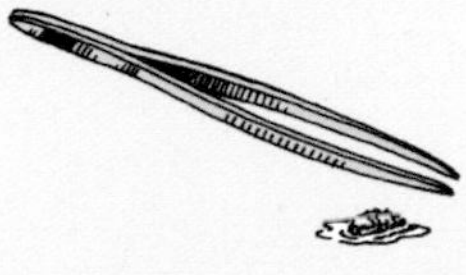

1 The larva needs to breathe, so cut off its air supply by covering it with duct tape.

2 Apply pressure around the wound and grasp the larva tail with tweezers when it comes out.

3 Pull until the larva is completely out. Clean and bandage the wound.

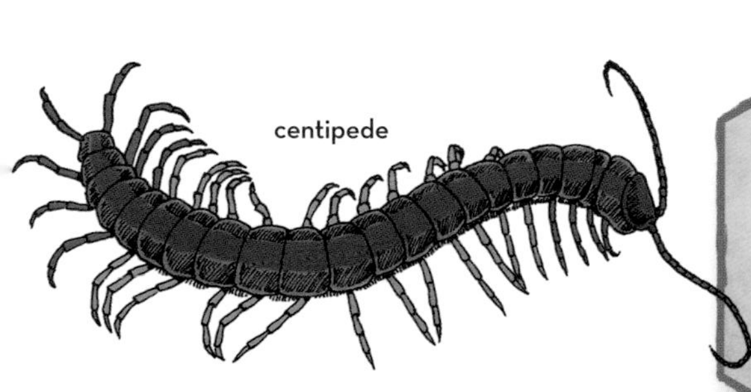
centipede

BEAR SAYS

Centipedes, especially the larger species, can inflict painful, venomous bites. If bitten, clean the wound and seek help.

Arachnids

Arachnids have eight legs and a body made up of two parts. These are some of the most dangerous kinds.

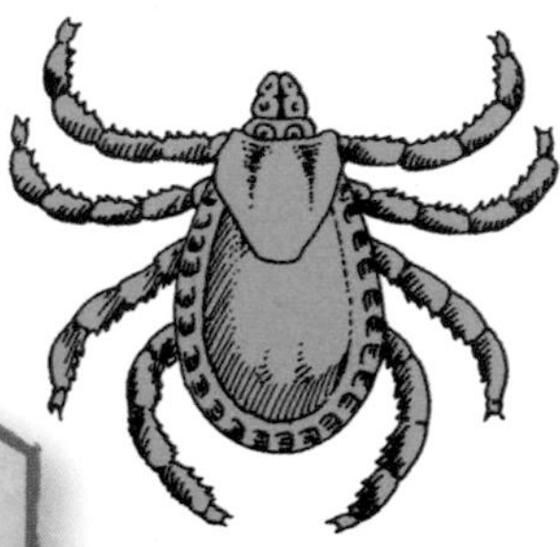

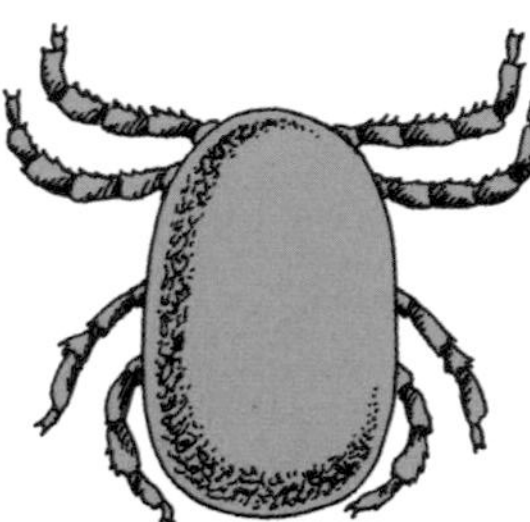

To remove a tick, use thin tweezers, and grasp the tick as close to the skin as possible. Pull upward with a steady, even pressure.

Hard tick

These tiny bloodsucking arachnids are responsible for the spread of many illnesses. The hard tick family comprises the majority of tick species. They have a hard shield-like plate just behind their mouthparts.

Soft tick

The less common soft ticks have a rounded, leathery appearance with mouthparts that can't be seen from above. They feed mostly on birds and small mammals, but will also choose human hosts.

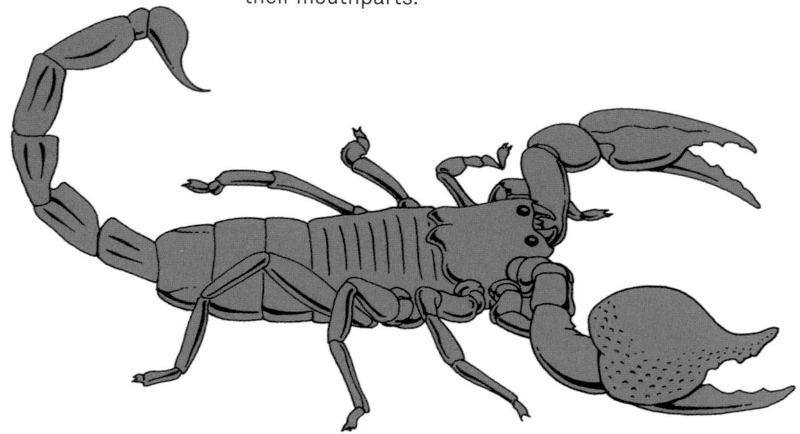

Scorpion

About 25 species of scorpions can kill. They live in northern Africa, the Middle East, India, Mexico, and parts of South America. Most of the other 1,000 or so species can deliver a very painful sting.

Funnel-web spider

There are about 40 species of funnel-web spiders in Australia. The highly venomous Sydney funnel-web spider is possibly the world's most dangerous spider. It is likely to strike repeatedly if disturbed.

Widow spider

Many spiders in this animal group are highly venomous. Well-known species include the black widow (North America), the redback spider (Australia) and button spiders (southern Africa). Bites can be deadly.

Brazilian wandering spider

This group of aggressive spiders is found in Central and South America, and in banana shipments worldwide. Their venom is the most toxic of any spider.

Tarantula

These frightening-looking spiders are actually quite timid. Most bites are similar to a wasp sting, although one species causes hallucinations. Some kinds shed irritating hairs as a form of defense.

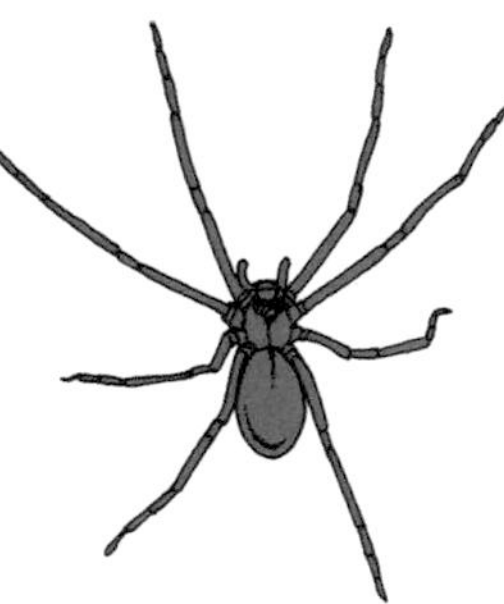

Recluse spider

Bites from these spiders can cause cell tissue death requiring skin grafts and other surgery in extreme cases. Mild skin damage and itchiness is more common, however.

Reptiles

Cold-blooded creatures, such as reptiles are covered with scales or bony plates. Snakes and lizards belong to this group—and many can deliver a potentially fatal bite.

Snakes of the Americas

Copperhead

These well-camouflaged North American snakes will often freeze when feeling threatened. This means that bites often happen when they are stepped on by accident. Luckily, their bites are rarely fatal.

Rattlesnake

These snakes cause the majority of snake injuries and deaths in North America (even so, deaths are very rare). Despite their deadly reputation, rattlesnakes are timid, normally giving a warning rattle when alarmed.

Bushmaster

This genus of large venomous vipers is found in remote forested areas of Central and South America. The bushmaster is capable of repeated strikes and the injection of large amounts of venom.

Coral snake

There are over 65 recognized species of coral snakes in the Americas. They have very potent venom, but because of their mild nature and small fangs, deaths and injuries are rare. Many harmless snakes mimic the coral snakes' coloration for protection.

Cottonmouth

This viper is native to the southeastern United States. A cottonmouth will vibrate its tail and throw its mouth open as a threat display. Bites are painful and can be fatal.

African and Asian snakes

Boomslang

The venom of this sub-Saharan snake works as a hemotoxin—even small amounts will cause severe internal and external bleeding. They will strike fast if disturbed.

Cobra

Most cobra species rear up and spread their necks in a threat display. Some can "spit" venom up to 8 ft. They aim for their attacker's eyes. A direct hit causes severe burning pain.

Krait

This group of snakes is found in the jungles of India and Southeast Asia. They are armed with a neurotoxin that causes muscle paralysis (loss of movement).

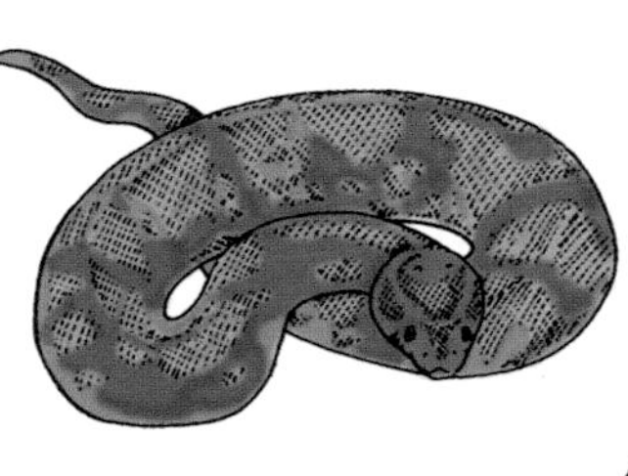

Saw-scaled viper

These small snakes live in dry savannah habitats. They make a rasping sound when alarmed by rubbing the sides of their bodies together. They are very dangerous.

Puff adder

This snake species is responsible for more snakebite deaths in Africa than any other. When approached, it draws its head close to its coils, makes a loud hissing sound and is quick to strike.

Mamba

Most mamba species are tree-dwelling. The exception is the land-based black mamba—the world's fastest, and Africa's deadliest, snake. Untreated, its bite is fatal.

Australian snakes

Eastern brown snake

This snake species is responsible for most deaths caused by snakebite in Australia. Its venom is the most toxic of any land snake in the world, except for the inland taipan.

Red-bellied black

The red-bellied black is commonly found in woodlands, forests, swamplands, and urban areas of eastern Australia. They usually avoid attack. Bites are dangerous but rarely fatal.

Taipan

All species in this group are dangerous. The inland taipan is viewed as the most venomous land snake in the world. However, the human population of its habitat is low, and all bite victims have been successfully treated with antivenin.

Tiger snake

The common tiger snake is found in southern and eastern Australia. Their highly toxic venom is produced in large amounts. The venom mainly affects the central nervous system, but also causes muscle damage, and affects blood clotting.

Death adder

Death adders are found in most parts of Australia, New Guinea, and nearby islands. They have relatively large fangs and toxic venom. Before the introduction of antivenin, about 60 percent of bites to humans were fatal.

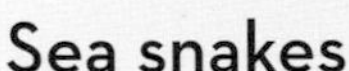

Sea snakes

Hydrophiinae

Found in warm coastal waters from the Indian Ocean to the Pacific, some species have venom more toxic than any land snake. Sea snakes are curious and will readily approach divers and swimmers, but they are generally placid and unlikely to attack.

European snakes

Adder

The common adder is the only poisonous snake of northern Europe. It is widespread in highly populated areas, and bites are fairly common—but very rarely fatal. The common adder has several larger and more dangerous relatives in southern Europe.

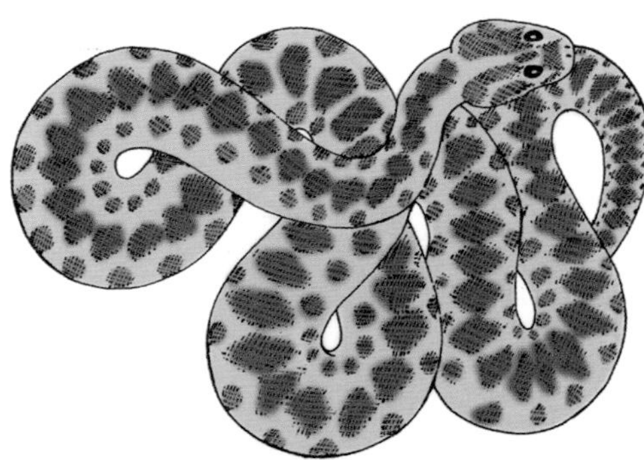

BEAR SAYS

Snake venom can be deadly. It can contain neurotoxins that affect the body's nervous system, or hemotoxins that destroy red blood cells. Stay away!

Avoiding snakebites

Snakes are timid creatures. Unless they are stepped on, cornered, or handled, they present very little danger to humans. When you are in their habitat, follow these tips to avoid a bite.

sturdy boots and gaiters will protect vulnerable body parts low to the ground

carry a stick and use it to push aside logs and shrubbery

keep to cleared tracks as much as possible

Treating snakebites

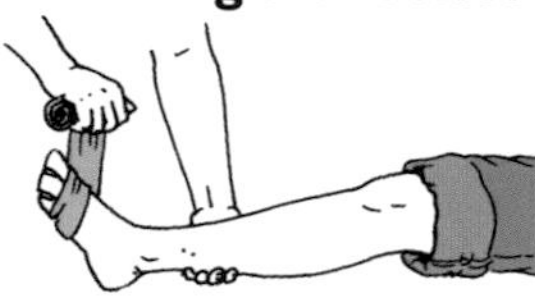

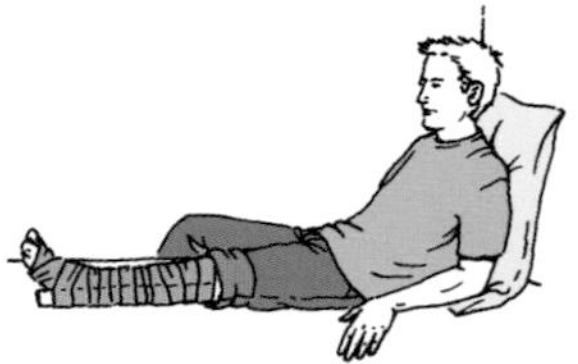

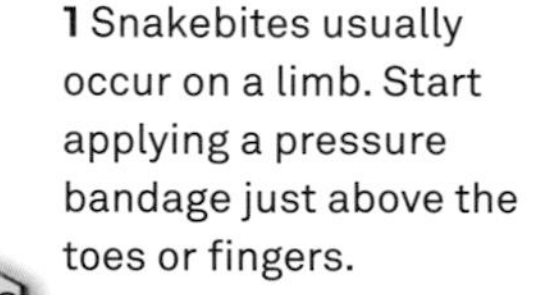

1 Snakebites usually occur on a limb. Start applying a pressure bandage just above the toes or fingers.

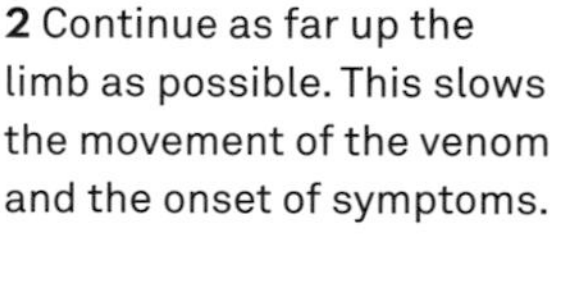

2 Continue as far up the limb as possible. This slows the movement of the venom and the onset of symptoms.

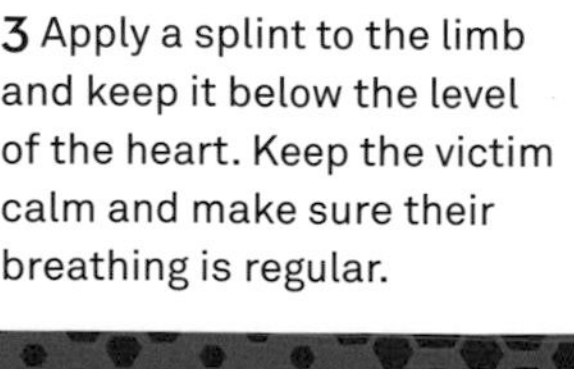

3 Apply a splint to the limb and keep it below the level of the heart. Keep the victim calm and make sure their breathing is regular.

Lizards

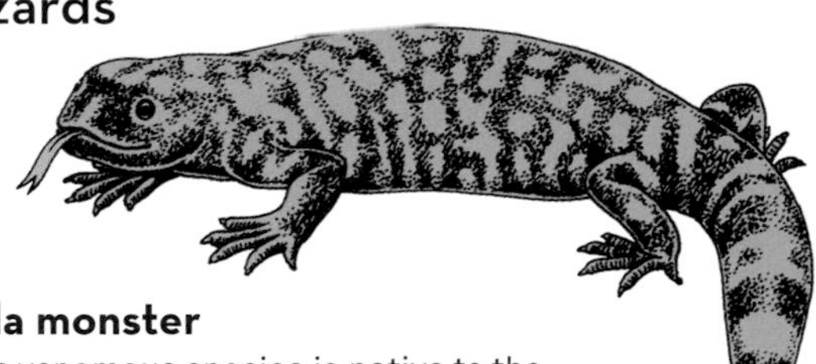

It is rare for a snake to actually chase a person, so if you come across one in the wild, stay calm and back away slowly.

Gila monster

This venomous species is native to the southwestern US and the northwestern Mexican state of Sonora. Although a Gila bite is extremely painful, none has resulted in a confirmed human death.

Beaded lizard

A close relative of the Gila monster, the beaded lizard is found mostly in Mexico and southern Guatemala. Its bite causes terrible pain, swelling, and a rapid drop in blood pressure.

Crocodile

Two crocodile species—the Nile crocodile and the saltwater crocodile—are man-eaters. Stay well away from water where they are known to be present.

Alligator

The American alligator is native to the southeastern US. Alligators occasionally attack unprovoked, and their bites can cause dangerous infections.

Aquatic animals

Our oceans, rivers, and lakes can be deadly places. Knowing and recognizing these creatures is highly important if you are spending time near water.

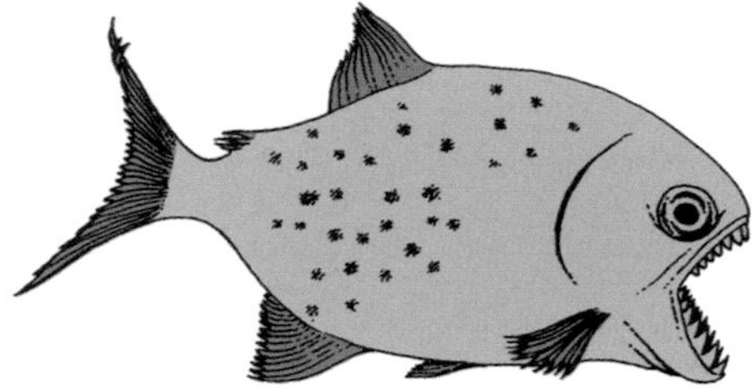

Piranha

The mouths of these South American freshwater fish are packed with sharp triangular teeth that can puncture and rip. While they will not strip humans to the bone, piranhas will take bites of flesh and remove toes.

Candiru

The Amazon's most feared fish usually survives by invading the gills of larger fish, where it feeds on blood. However, it has also been known to lodge itself in the human urethra (the tube that connects the bladder to the outside of the body).

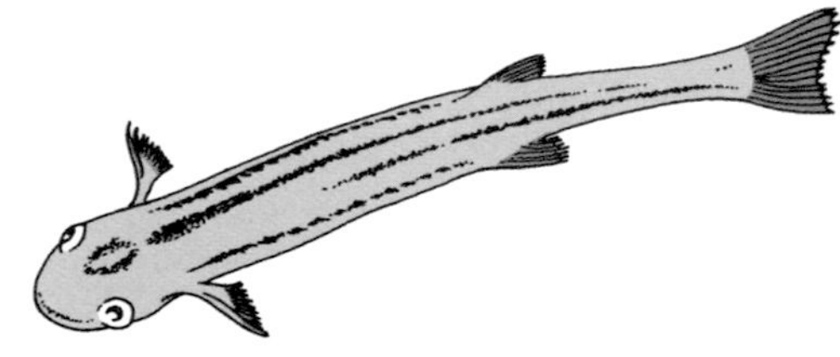

Bull shark

This shark species lives in both the open ocean and hundreds of miles up rivers. Because of their wide habitat range and aggression, many experts consider them the most dangerous shark species.

Electric eel

When angry, these large South American fish can deliver a burst of 600 volts—more than enough to kill. However, such deaths are very rare.

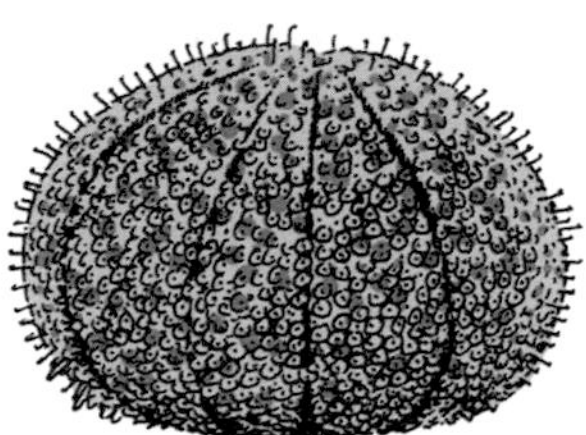

Flower urchin

Many sea urchins are armed with sharp spines and should be avoided. The spines of a flower urchin inject an extremely toxic venom. Injuries are very painful, and deaths have been reported.

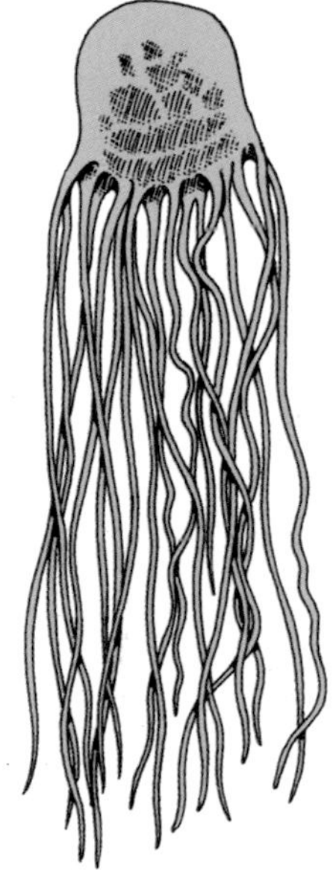

Box jellyfish

These deadly jellyfish live in coastal waters off northern Australia and throughout the Indo-Pacific. A box jellyfish sting is so excruciating and overwhelming that a victim can go into shock and drown if swimming alone. Heart failure often follows.

Portuguese man-of-war

The sting of the Portuguese man-of-war causes severe pain and in some cases, fever and shock as well as heart and breathing problems. To treat, remove any stingers that are still attached, wash with seawater, then submerge the affected area in hot water.

Cone shell

These pretty marine snails use unique venoms to hunt their prey. A sting from a large cone shell brings severe pain and is potentially fatal. Treat as though it is a snakebite—there is no antivenin cure.

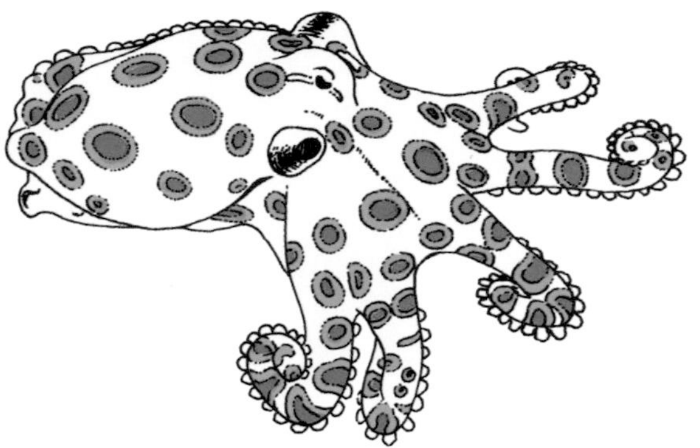

Blue-ringed octopus

These octopus live in tide pools in the Pacific Ocean from Japan to Australia. Although small and docile, they carry enough venom to kill 26 adults within minutes. Stings can bring total paralysis without loss of consciousness. Victims require artificial respiration for survival.

Needle fish

These shallow marine-dwelling fish make short jumps out of the water at speeds up to 40 mph. Their sharp beaks can inflict deep wounds and often break off inside the victim.

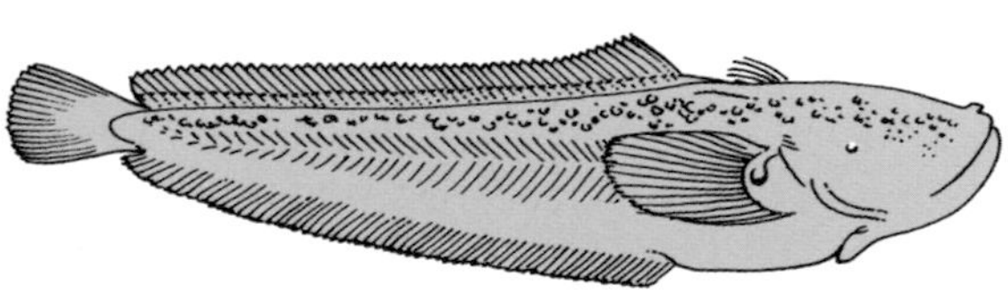

Toadfish

Venomous toadfish dwell in tropical waters off the coasts of Central and South America. They have very sharp, very poisonous spines hidden in their dorsal fins. They bury themselves in the sand and are easily stepped on.

Stonefish

The world's most venomous fish live in the coastal waters of the tropical Indo-Pacific. Symptoms of its venom are muscle weakness, temporary paralysis, and shock, which may cause death if not treated.

Shark

Although greatly feared, shark attacks on humans are extremely rare. No species is thought to target people as prey. Only a few kinds (the great white, tiger, and bull shark) have been involved in a significant number of fatal, unprovoked attacks.

Lionfish

This genus of aggressive fish is native to the tropical Indo-Pacific and has been introduced into the Atlantic coastal waters of the United States. Its venom can result in vomiting, fever, and sweating, and has been lethal in a few cases.

Stingray

These fish are mostly gentle, yet have a venomous barbed stinger on the tail. People are usually stung accidentally when stepping on a stingray. Stings can result in pain, swelling, nausea, and muscle cramps.

Mammals

Mammals are warm-blooded animals that have fur or hair on their bodies, and they feed their babies with milk. Some large mammals, such as these, can be particularly threatening if disturbed.

American black bear

These medium-sized bears rarely attack humans, but you should still avoid contact if possible. The most dangerous black bears are those that are hungry or have become used to human contact.

Brown bear

These large bears are normally unpredictable, and will attack if they are surprised or feel threatened. Mothers with cubs are particularly dangerous. If attacked, protect the back of the neck and play dead.

Polar bear

Contact with the world's largest land carnivore, or meat eater, should be avoided. A well-fed polar bear may show signs of curiosity near humans, while a hungry bear may stalk, kill, and eat you. Escape is unlikely without a weapon, but you could curl up and play dead.

Vampire bat

The common vampire bat is native to the Americas' tropics and subtropics. They will feed on human blood when horse and cattle are in short supply. Their bites can cause rabies, a deadly viral infection.

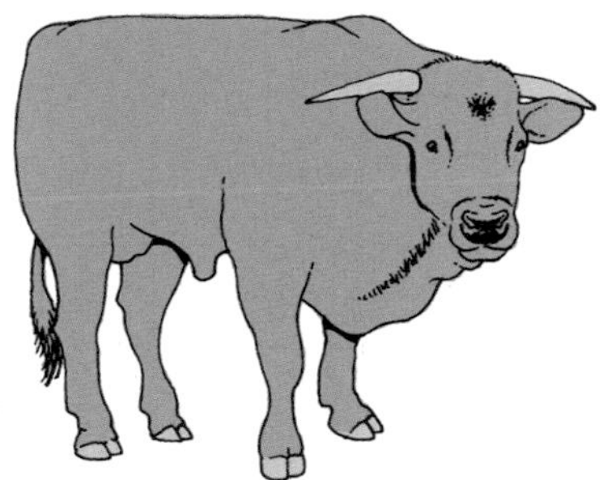

Bull

There is nothing like a large bull to turn a pleasant walk into a frightening ordeal. Never turn your back on a bull that has its head lowered or is pawing the ground. Back away slowly.

Rhinoceros

The five living species of rhinoceros are known for charging without being provoked. With very poor eyesight, they often panic at unusual smells and sounds.

BEAR SAYS

Never anger a hippopotamus! They are responsible for more human deaths in Africa than any other large animal.

Tiger

The tiger is the largest of the cat species. Human prey appears to be a last resort for tigers, but individual tigers have been responsible for the deaths of hundreds of people.

Lion

As with tigers, humans are not a favored prey of lions. However, where human settlements encroach on lion territory and regular prey animals are in short supply, lions will hunt and kill humans.

Leopard

Attacks by leopards on humans are rare, however injured, sick, or struggling individuals may turn to human flesh. The "Leopard of Panar" is reported to have killed as many as 400 people in northern India in the early years of the twentieth century.

Wolf

Like any large predator, a wolf is potentially dangerous, and common sense tells us to avoid them. Fortunately, attacks on people are very rare. Wolves with the disease rabies, in the "furious" stage, are the most dangerous.

DANGEROUS PLANTS

Plants have evolved a range of effective defenses against animals that might want to eat them. Some need just the lightest touch and you're in trouble.

Stinging nettle

This plant is common in many temperate parts of the world. Hairs on the leaves and stems contain irritating chemicals, which are released when the plant comes into contact with skin.

Cacti

Large cactus spines can be removed with tweezers. Work slowly because some spines have barbed ends. To remove very small, fine spines, apply duct tape to the area, then gently remove it.

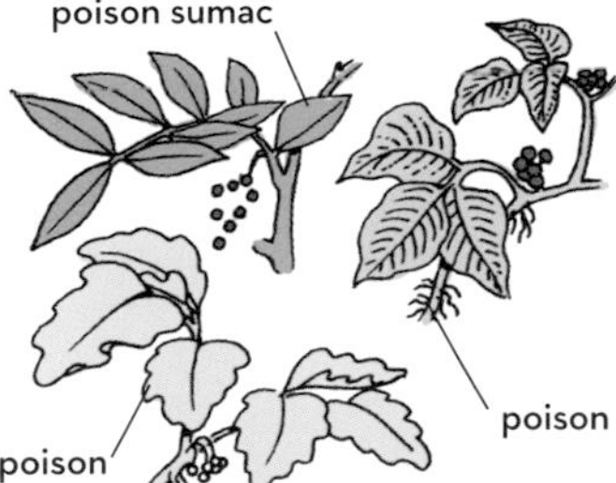

Poison ivy, poison oak, and poison sumac

These plants grow in parts of the United States and southern Canada. People react differently if they come into contact with them, but most will be affected by rashes and blistering.

Stinging trees

There are about 37 species of stinging trees across Southeast Asia, Australia, and the Pacific Islands. Avoid the Australian Gympie-Gympie, as one touch can mean months of agonizing pain.

QUICKSAND

Quicksand is a mass of fine sand, silt, and clay that has become completely saturated with water. While it's hard to get out of, it is possible to escape.

1 If you feel yourself sinking into quicksand, act fast.

2 Unstrap your pack or any other heavy gear and throw it aside.

3 Drop onto your back to spread your weight. Then, work to free your legs.

4 Use swimming or snake-like motions to return to solid ground. It may take hours to move a few feet, but you can take a break at any time.

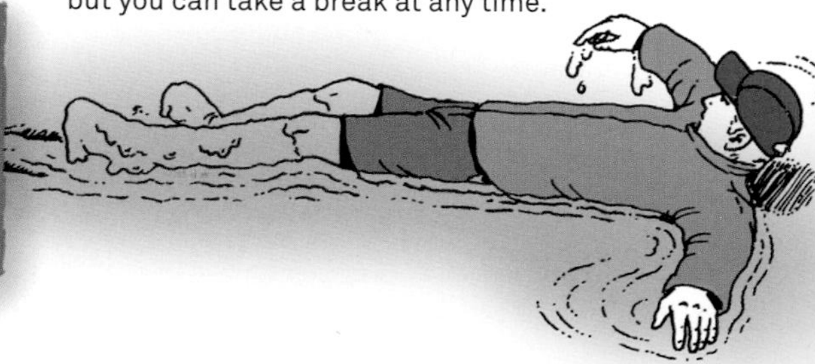

BEAR SAYS

Escaping from quicksand is a battle. The number one rule is to keep calm. Once you have escaped, clean yourself off to prevent further harm from chafing.

SOURCING WATER

If you are lost or your supplies are running low, your first task should be to find water. In some places water is easily found, but in arid (very dry) areas it can be a life-or-death challenge.

Animal indicators

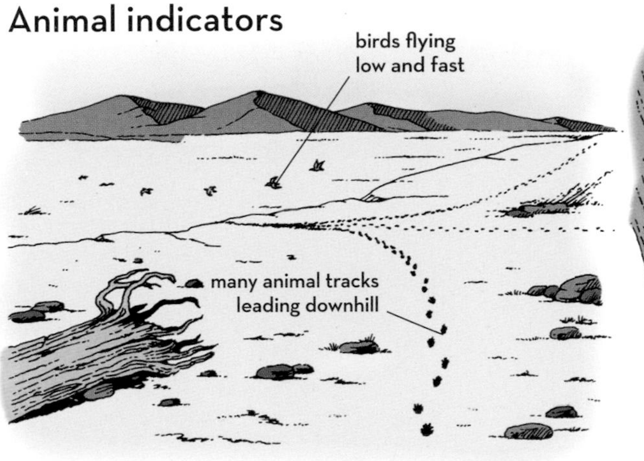

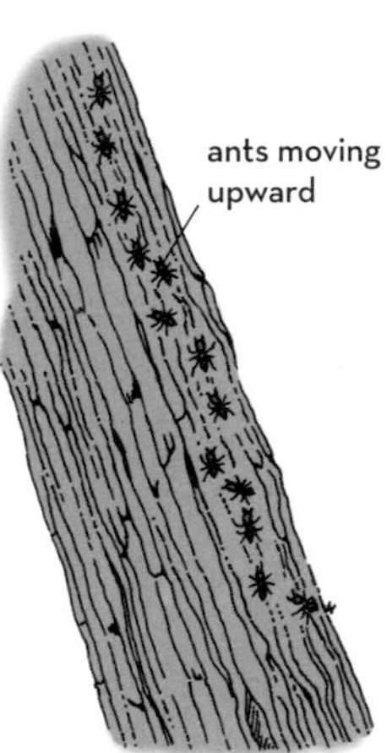

Follow the animals

All living things need water to survive. Observe the animals in your surroundings and you will get valuable clues. Don't forget to look for plants. Unusually green plants in an arid landscape may indicate water just below the surface.

Ants in a row

A column of ants heading up a tree trunk may be heading to a reservoir of water.

Distance to water

Bees

Usually within 3 miles.

Flies

Usually within 1.5 miles.

Mosquitoes

Usually within 1,500 ft.

Frogs

Usually in the immediate area.

BEAR SAYS

Don't give up too soon! You may need to dig down 3 ft. or more to reach water.

Dry riverbed

In a waterless landscape, a dry, sandy riverbed is often the best place to look for water. The best places to dig are the lowest points, the outside of bends, and near where green plants are growing.

Cliff base

Water naturally pools at the base of cliffs and hills. Such pools are deep and often the last to disappear because they are protected or partially protected from the sun. If no water is found, dig in places where it would pool after rain.

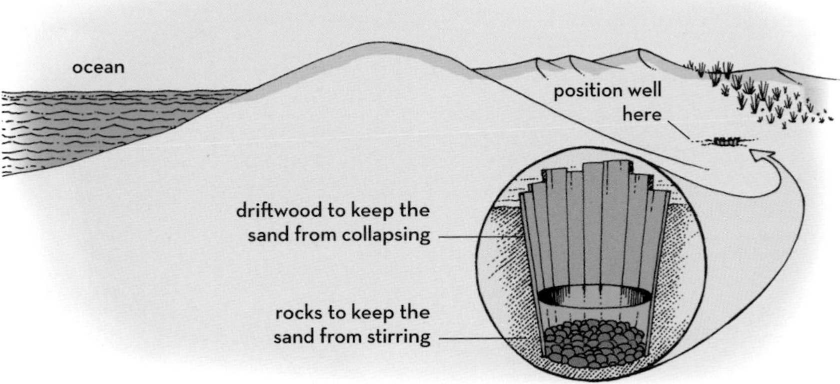

Beach

A beach well is just a hole, dug behind the very first sand dune in from the ocean. It should be about 3 ft. deep. Fresh groundwater seeping toward the ocean will gather in the well and float on top of the salty seawater.

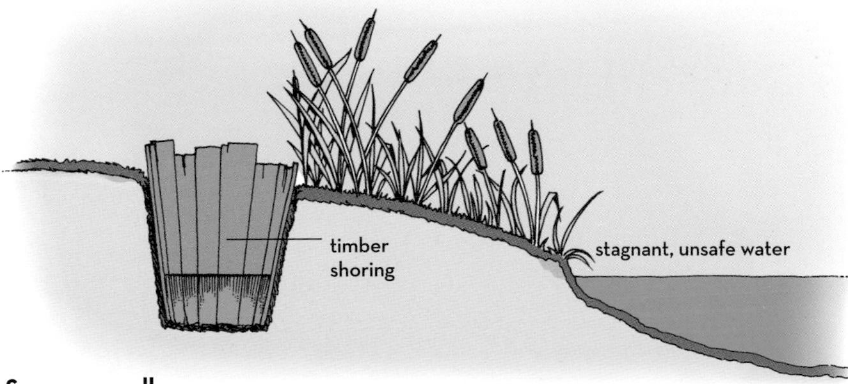

Seepage well

You may come across a stagnant body of water that is cloudy, has a bad flavor or odor, or is difficult to access. If this happens, dig a well about 30 ft. from the water source. The water that fills the well will be filtered and should be safe to drink.

WATER FROM PLANTS

If you can't find a water source in the environment around you, you can search for certain plants that can provide a drink.

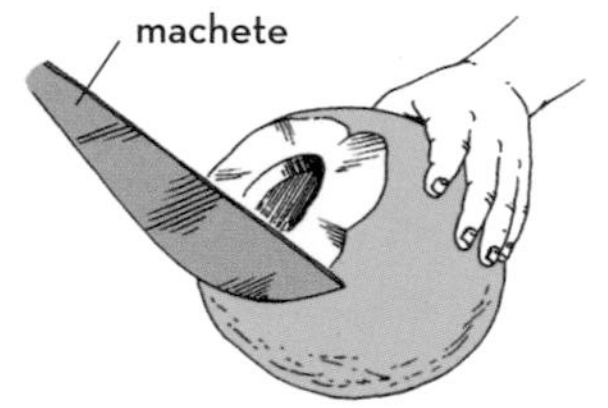

Mature coconut

Use a sharpened stake driven into the ground to split and remove the outer husk and reveal the shell. Drive a hole through a soft "eye" of the shell to access the coconut water.

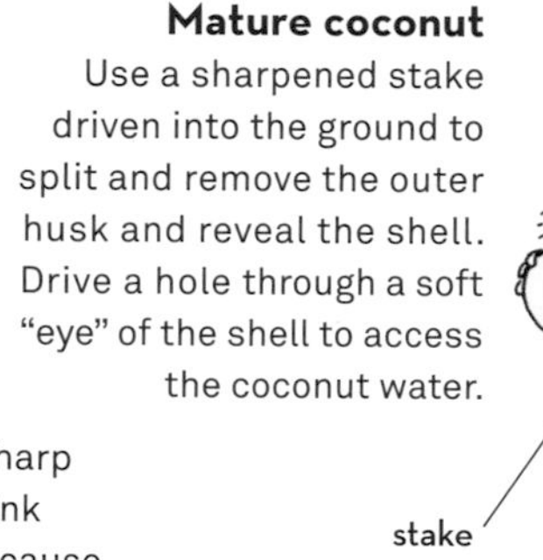

Green coconut

Slice open a green coconut with a sharp knife to access the water inside. Drink the coconut water in moderation because it is a natural laxative.

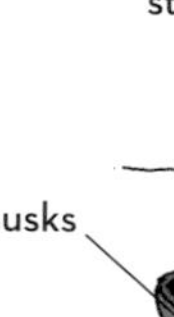

Banana tree

Cut through the trunk of a banana tree about 4 in. above the ground. Then, hollow out a bowl-like reservoir inside the stump. Water from the roots will gather in the bowl. Scoop the water out of the bowl three times before drinking as the water will be bitter at first.

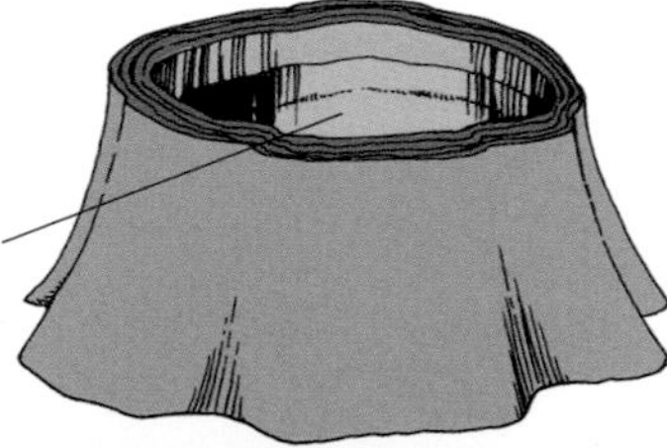

Green bamboo

Green bamboo can supply you with fresh water, even at the height of the tropical dry season. To collect water from a young stalk, bend it over, tie it securely, and cut off the top. Water will drip out of the cut. Collect it in a container.

green bamboo shoot

collection container

stake and rope

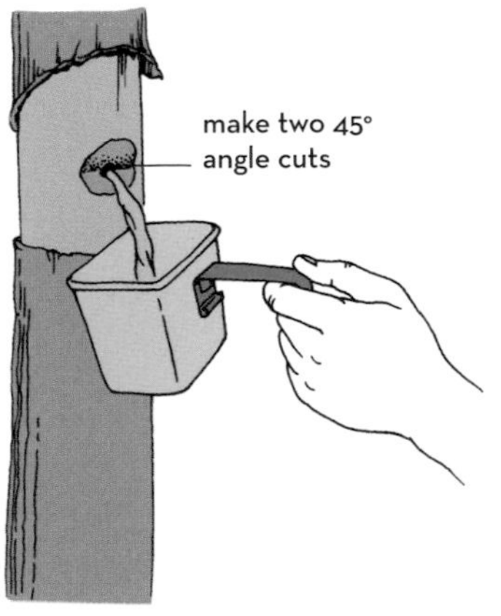

Water from a vine

Cut a section of a vine high up. Then, sever it completely near the ground. Liquid will drain out the bottom. Don't drink from vines that produce white sap or milky liquid when cut. Get rid of liquid that has a sour or bitter taste.

Big bamboo

For larger shoots of green bamboo, simply cut a hole near the base of each section and collect the water within.

Transpiration

This is one of the most efficient and easily constructed sources of water in an arid setting. Tie a plastic bag around a leafy branch of a medium-sized tree or shrub, and place a container underneath. After a few hours in the sun, you will have some clean, drinkable water.

BEAR SAYS

Finding clean, safe water should always be your top priority. You can survive three weeks without eating, but only three days without water, so find it fast.

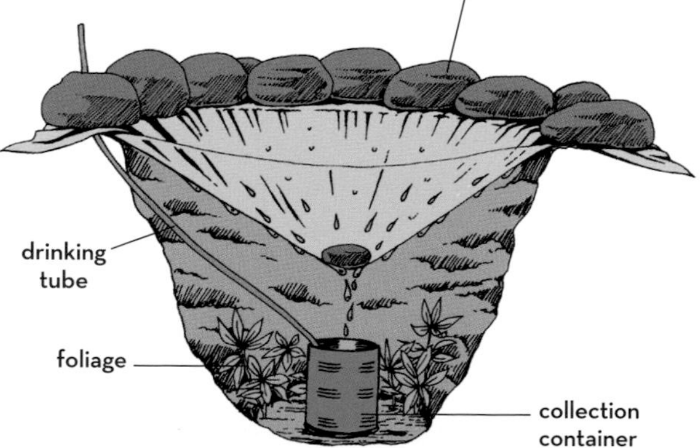

Solar still

This system extracts water from the soil and any plants growing in or placed in a hole. Moisture evaporates, rises, and then condenses on the underside of the plastic barrier above, which then drips into a collection container below.

Water from cuttings

Collect as many green leaves and branches as can fit in a plastic bag without touching the sides. Prop up the center to form a tent. Arrange the bag on a slight slope so the condensation will run down to a collection point.

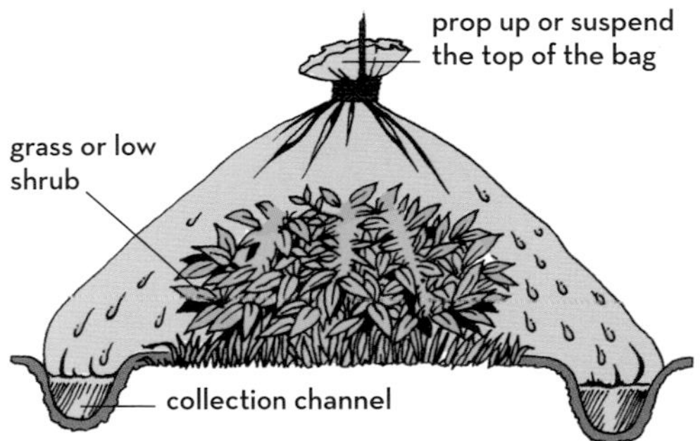

Ground transpiration

This arrangement works on the same principle as the transpiration bag opposite, but with plants that grow close to the ground. Consider scaling up from a plastic bag by using the fly of a tent.

Cacti

Cacti are a valuable survival resource in many deserts. The fruits of the prickly pear and some other species are edible. Many cacti contain huge amounts of water in their flesh that can be gathered in solar stills or transpiration bags. Cacti are protected in some areas and should only be used in an emergency.

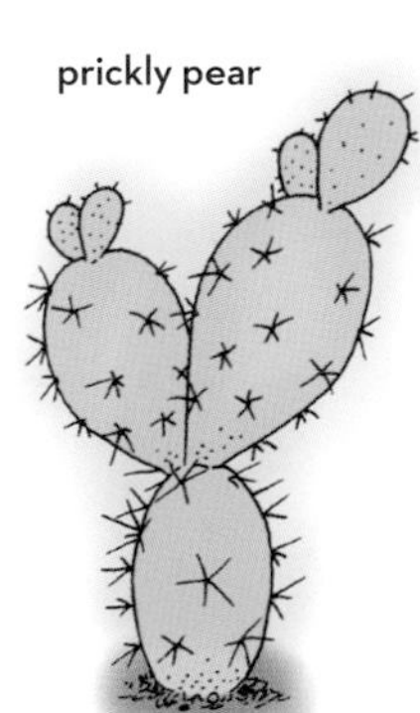

WATER PURIFICATION

In the wild, even water that looks pure and pristine may not be. Luckily, having clean drinking water is relatively simple when you have the right equipment and knowledge.

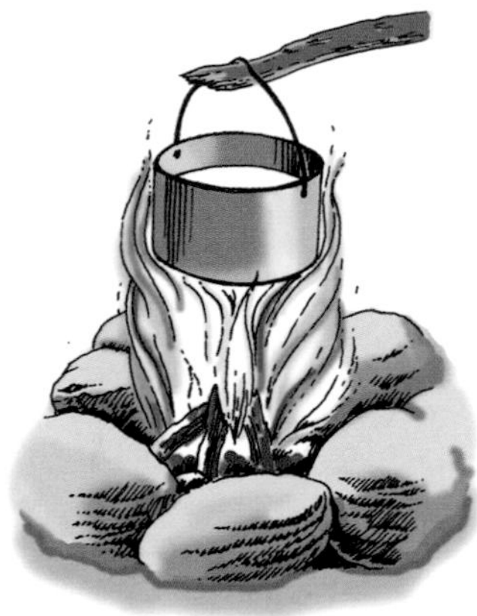

Boil

This kills most types of disease-causing organisms. Boil the water for at least one minute, then let it cool down.

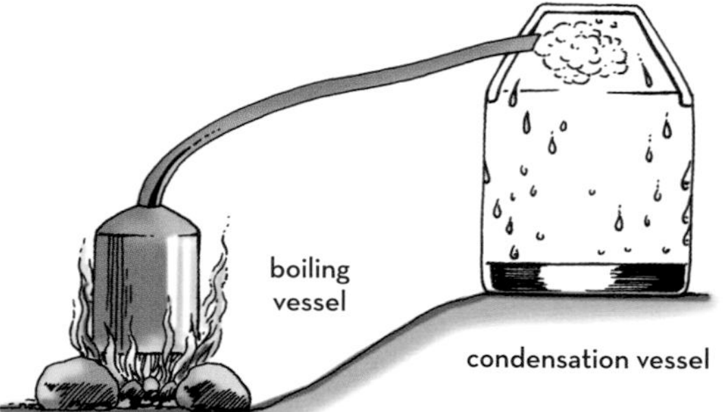

Distill

This system is trickier than simple boiling, but it makes drinkable water from sources heavy with sediment. It can also be used to distill seawater or urine.

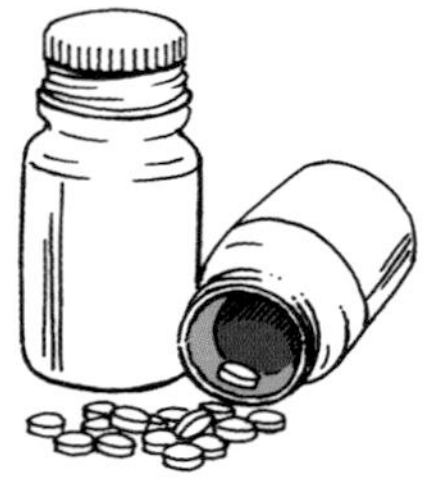

Chemical

Iodine, potassium permanganate, and chlorine can be used to treat water. They take time to work, and be prepared for a slightly odd taste.

Solar

Pour suspect water into clear plastic bottles and expose them to direct sunlight for at least six hours (or for two days in very cloudy conditions).

Filters

Thorough water filtration removes particles and many microorganisms that cause disease, but it's still a good idea to boil the water before drinking it.

BEAR SAYS

If you don't have a filter, you can make one using a plastic bottle, gravel, sand, and charcoal. The charcoal helps to filter out tiny impurities in the water.

virus and bacteria filter

parasite filter

carbon filter

Bottle filter

This filter bottle works with a cartridge that needs to be replaced after every 150 or so refills. Simply squeeze to produce a flow of water.

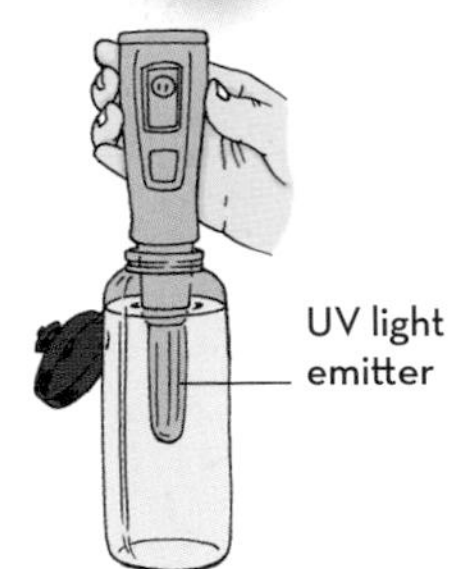

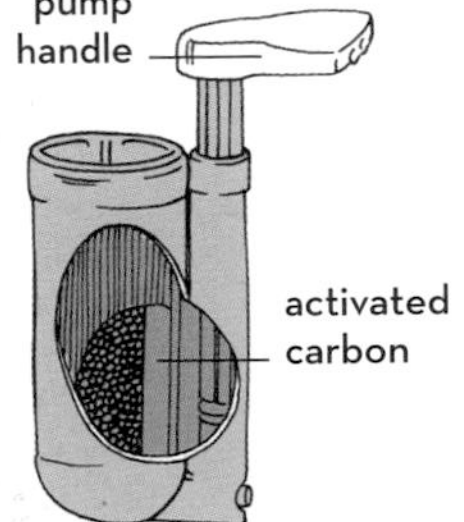

Pump filter

There are many pump filtration systems that hikers can use. Each stroke of the pump draws water through the filter and purifies it.

Ultraviolet

This battery-driven device uses UV light to sterilize 33 oz. of water in 90 seconds. The water must be clear for the sterilization to work well.

FOOD FROM PLANTS

So you're stuck in the wilderness, and you've got plenty of safe drinking water. Your next priorities will probably be food and shelter. Plants can provide great nutrition if you know what's safe to eat.

Edibility test

1 Crush and smell the plant sample. Reject it if you sense strong, acid, or almond odors.

2 Crush and rub the sample against the inside of your elbow. Wait 15 minutes and discard if there is any irritation.

3 Hold a small amount against your lips. Reject if there is any irritation.

4 Place a small amount on the tongue. If there is any bad taste or irritation, throw it away.

5 Chew a small amount for several minutes, but do not swallow. If there is any irritation, spit it out.

6 If the plant part passes all these tests, eat a small amount and wait several hours for any adverse reaction.

BEAR SAYS

Tap along a piece of bamboo and listen to the noise it makes. Sections that have water inside will make a denser sound.

Split

Separate the plant into its basic components and test separately.

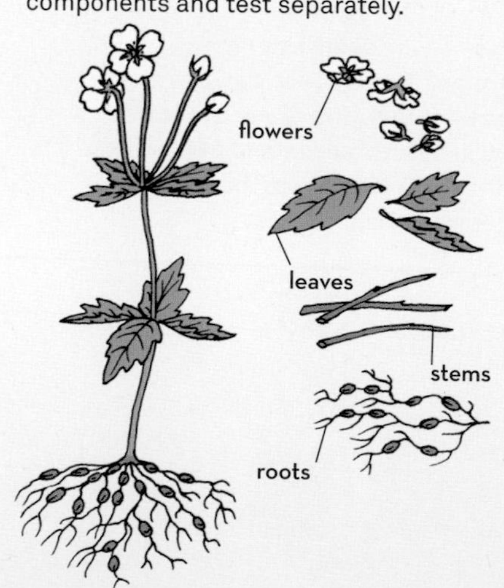

Plants to avoid

Some plants should be avoided altogether. Look for these indicators and leave them alone.

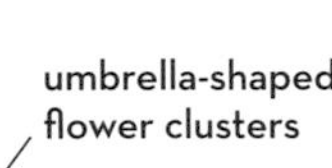

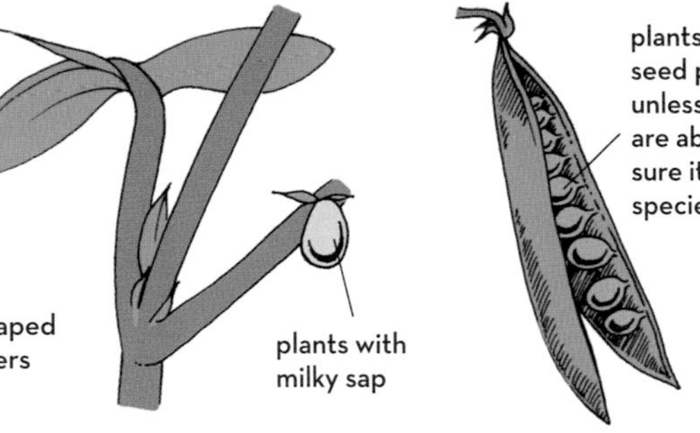

Plants to eat?

Most foods eaten by primates and birds are safe for us to eat, too. But this is not a guarantee—always use the edibility test opposite. Many berries are safe to eat, but only take a bite if you are certain they are edible, such as blackberries.

HUNTING SMALL ANIMALS

In a survival situation, a meal of meat goes a lot further than plants alone. While large animals can be difficult and dangerous to hunt, their smaller relatives are easier to get onto your plate.

Hand weapons
These are some tools traditionally used to catch small animals.

BEAR SAYS

Insects, frogs, lizards, and snakes are good sources of protein. Keep tiredness at bay with these energy sources in an emergency.

Catch a frog

Hold one hand about 1.5 ft. in front of the frog and slowly wriggle your fingers. This will grab the frog's attention. Grasp the frog from behind with your other hand.

stout stick

Catch a lizard

Gently wave a noose of tight wire in front of the lizard. Gradually bring the noose closer and closer, then lasso the lizard.

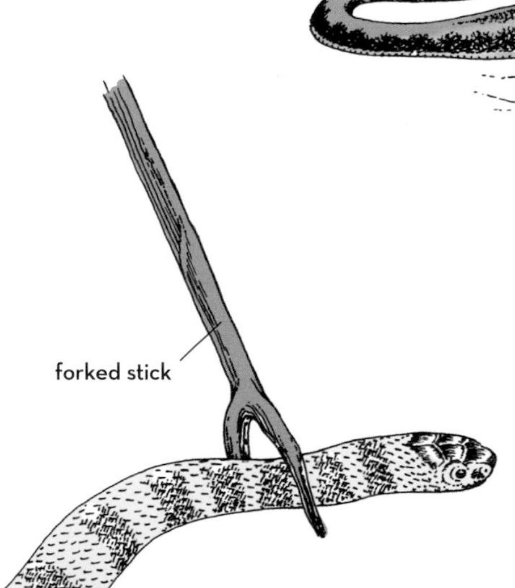

Catch a snake

All snakes can be eaten. To catch a snake, first stun it with a rock or stick. Pin its head down with a forked stick, and kill it with a knife, rock, or another stick. Cut the head off and bury it if there is any chance that it might be a venomous species.

Snares

These traps are used to catch some animals, often using wire or cord.

Simple loop snare

Make a snare about the size of your fist for small animals, such as rabbits. Set up about 15 snares for each animal you hope to catch.

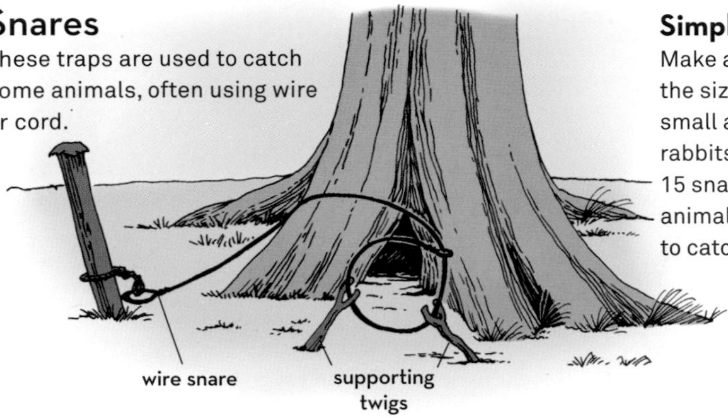
wire snare
supporting twigs

Spring snare

Game running through a spring snare releases the trigger bar and is flung off the ground. It's useful for catching rabbits, foxes, and similar animals.

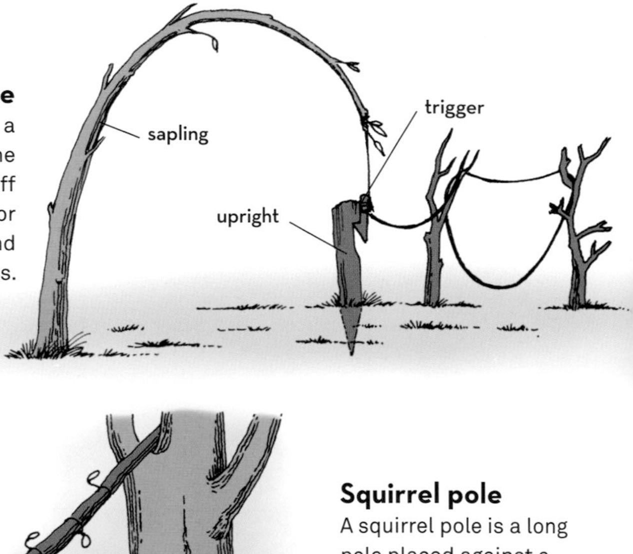
sapling
trigger
upright

close-up of snare

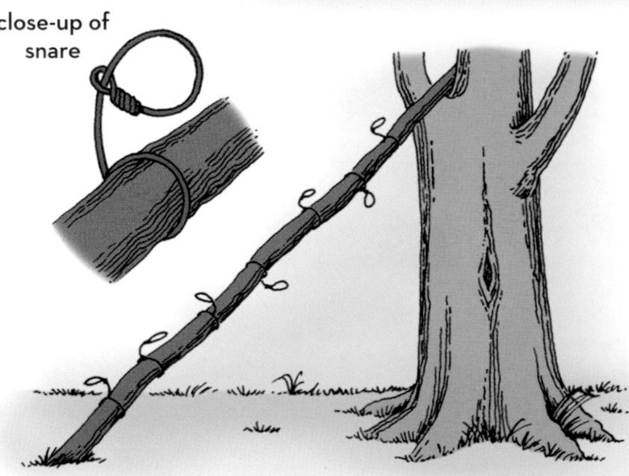

Squirrel pole

A squirrel pole is a long pole placed against a tree. Place several wire nooses along the top and sides of the pole so that a squirrel trying to go up or down the pole will have to pass through one or more.

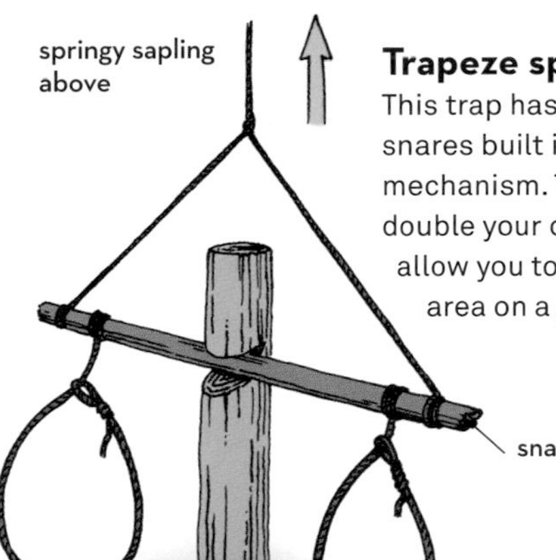

Trapeze spring snare

This trap has two separate snares built into one mechanism. This will double your chances, or allow you to cover more area on a game trail.

BEAR SAYS

In my experience it is always best to set as many traps as possible so that you have a greater chance of catching a meal.

Figure-4 deadfall trap

This simple trap can be made to any size. A horizontal bait bar is balanced at right angles to an upright with a lock bar, which supports a rock or other heavy weight.

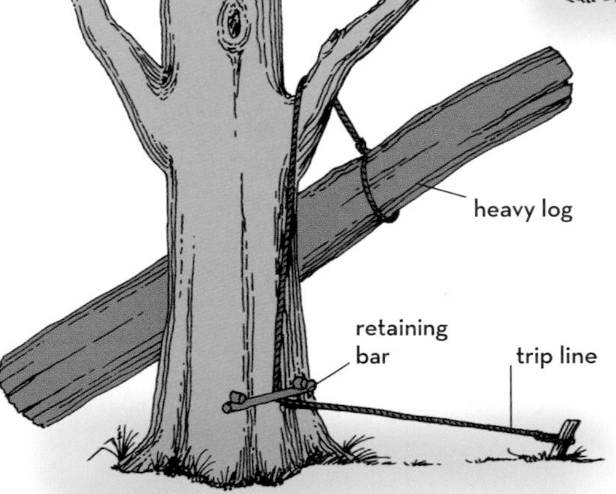

Deadfall trap

It's easy to build a deadfall trap large enough to kill a pig or deer. Make sure everyone in the party knows exactly where such a trap has been set as it could also harm a person.

EDIBLE INVERTEBRATES

Insects, mollusks, and arachnids can be found in large quantities and they are highly nutritious. If survival is at stake, put your taste buds aside and add some of these critters to the menu.

Worms

There are few better sources of protein than worms. Drop them in drinkable water after collection and they will naturally wash themselves out. If you prefer, dry and grind the worms and add them to soup.

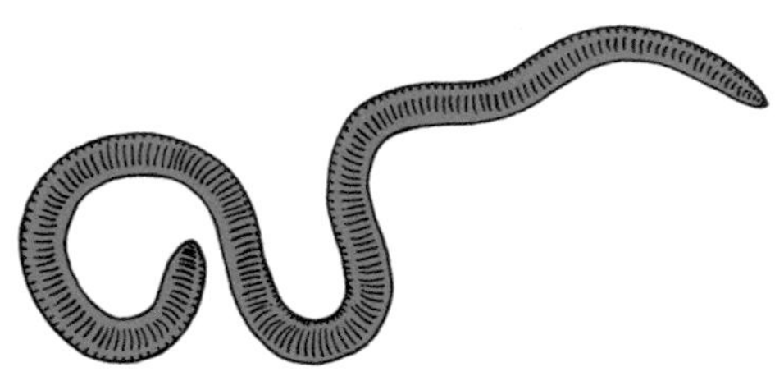

Snails

Starve snails for a few days so they can expel any poisonous plants they have eaten. Boil them for three minutes then drain, rinse in cold water, and remove from their shells. Avoid species with brightly colored shells.

Spiders

Don't overlook spiders as a source of protein. Eat the bodies and leave the heads, which may contain poison. If you catch a tarantula, try frying it—they are a delicacy in parts of Southeast Asia.

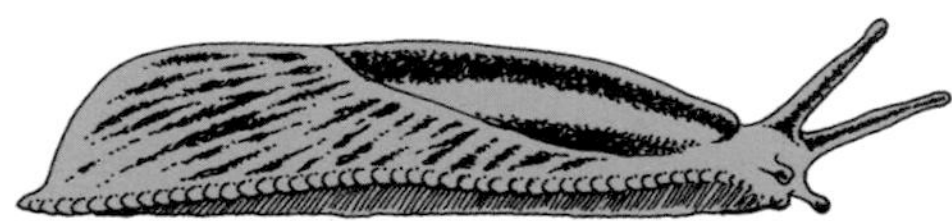

Slugs

Some slugs are very large—three or four will constitute a good meal. They can be eaten raw, but are much more appetizing cooked. Prepare and cook them exactly the same way as snails.

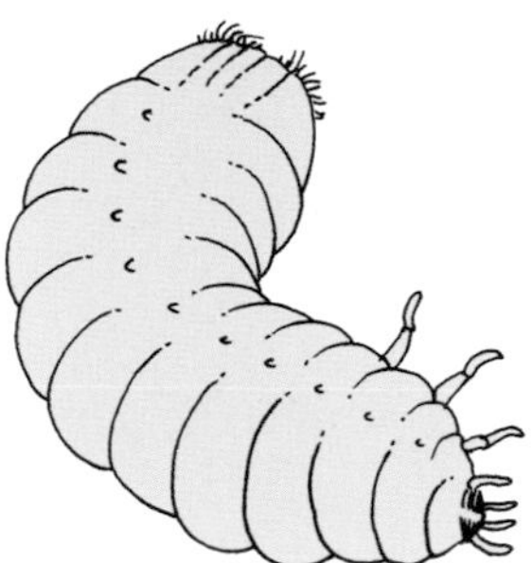

Grubs

Insect larvae, also known as grubs, are prime wilderness food. They favor cool, damp places, so look in rotten logs, under the bark of dead trees, under rocks, and in the ground. Grubs are safe to eat raw.

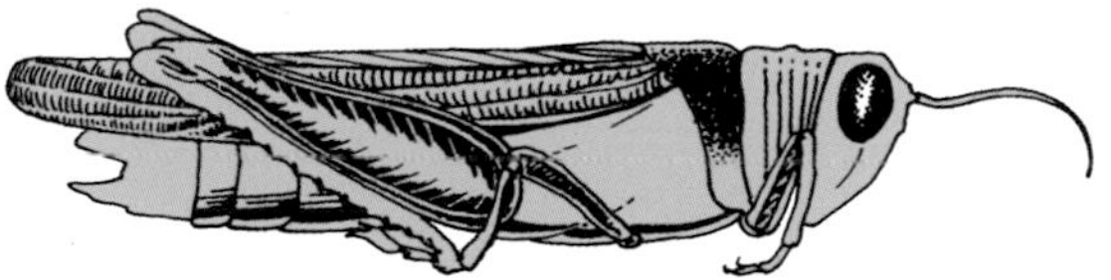

Grasshoppers

These insects can be a great source of food in some places. Knock them from the air with a piece of clothing or a leafy branch. Remove the wings, antennae, and legs before eating them. It is best to roast them to kill off any parasites.

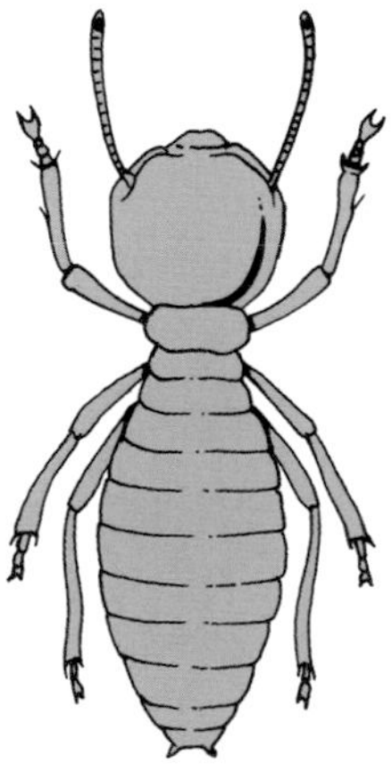

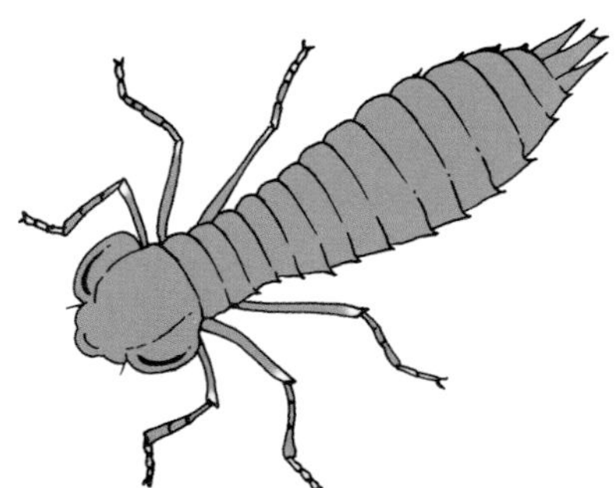

Aquatic insects

Nearly all water-based insects can be eaten in both adult and larval form. Use clothing as a net and trawl freshwater rivers or ponds.

Termites

These insects exist in enormous numbers in the warmer parts of the world and are easily collected from their nests. Remove the wings from larger species before eating. They can be cooked, but are more nutritious eaten raw.

EMERGENCY SHELTERS

Hot or cold, wet or dry, a good shelter is vital for your safety. Each landscape requires different types of shelter. Use the natural resources at hand depending on your need.

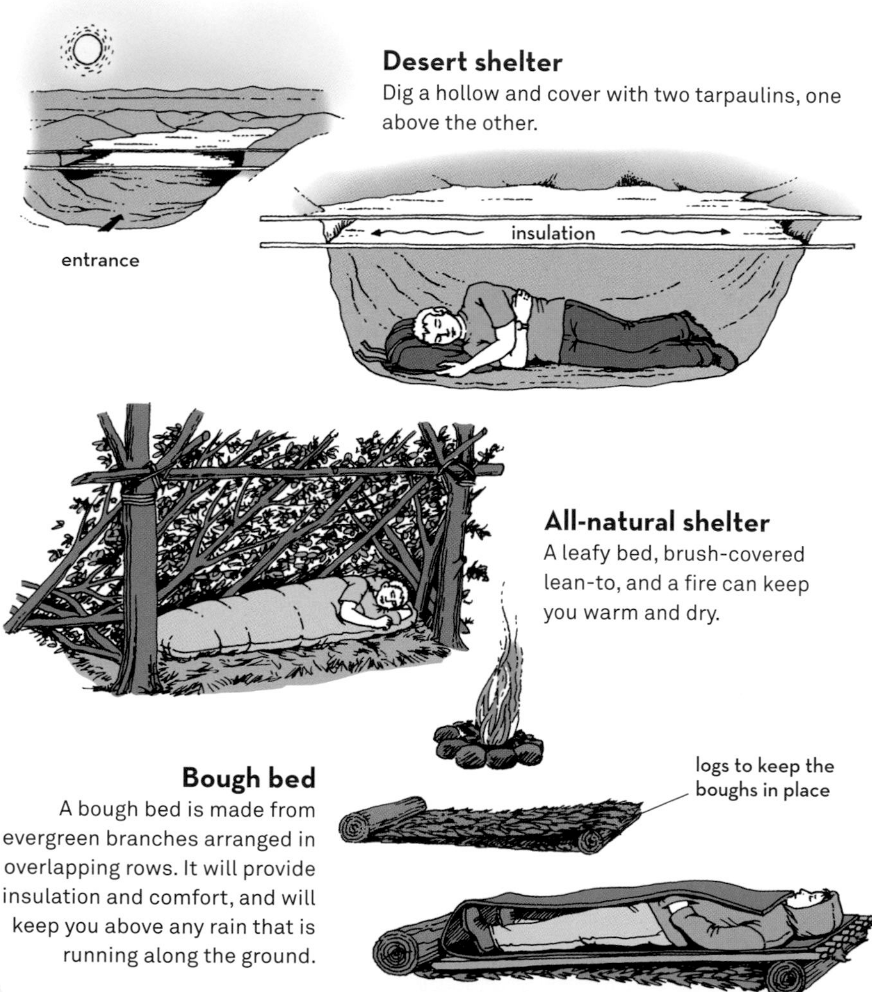

Desert shelter

Dig a hollow and cover with two tarpaulins, one above the other.

All-natural shelter

A leafy bed, brush-covered lean-to, and a fire can keep you warm and dry.

Bough bed

A bough bed is made from evergreen branches arranged in overlapping rows. It will provide insulation and comfort, and will keep you above any rain that is running along the ground.

Fallen tree shelter

A fallen tree can make a quick shelter. Improve it by removing branches on the underside and slinging a tarpaulin on top.

Sapling shelter

If you come across a group of saplings, clear the ground between them, strip their branches, and tie their tops together. Cover with material or weave branches between them.

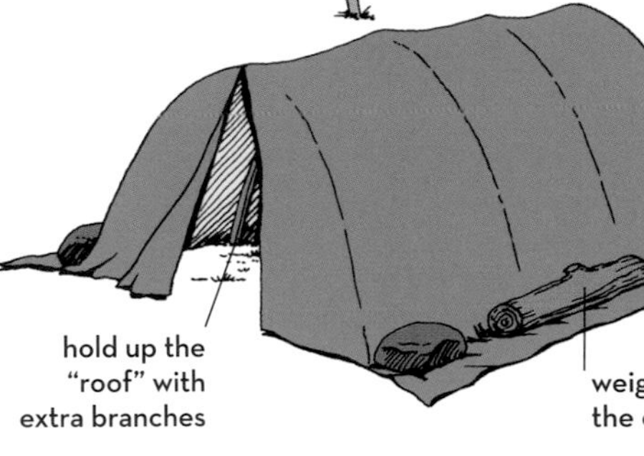

BEAR SAYS

In harsh weather or desperate situations, shelters such as these can save your life!

Tree-pit snow shelter

In forests where heavy snow has fallen, you will often find deep hollows under the branches of evergreen trees. Dig out some extra room if needed, and lay branches on the ground.

INDEX

Discover more amazing books in the Bear Grylls series:

Perfect for young adventurers, the *Survival Skills Handbooks* accompany an exciting range of reference, fiction, and coloring and activity books. Curious kids can also learn tips and tricks for almost any extreme situation in *Survival Camp*, and explore Earth in *Extreme Planet*.

First American Edition 2017
Kane Miller, A Division of EDC Publishing

Conceived by Bonnier Books UK in partnership with Bear Grylls Ventures
Produced by Bonnier Books UK, Suite 3.08 The Plaza, 535 King's Road, London SW10 0SZ, UK

For information contact:
Kane Miller, A Division of EDC Publishing
PO Box 470663
Tulsa, OK 74147-0663
www.kanemiller.com
www.edcpub.com
www.usbornebooksandmore.com

Library of Congress Control Number: 2017945577

Printed in Malaysia
5 6 7 8 9 10

ISBN: 978-1-61067-762-2

Disclaimer
Weldon Owen and Bear Grylls take pride in doing their best to get the facts right in putting together the information in this book, but occasionally something slips past their beady eyes. Therefore we make no warranties about the accuracy or completeness of the information in the book and to the maximum extent permitted, we disclaim all liability. Wherever possible, we will endeavor to correct any errors of fact at reprint.

Kids—if you want to try any of the activities in this book, please ask your parents first! Parents—all outdoor activities carry some degree of risk and we recommend that anyone participating in these activities be aware of the risks involved and seek professional instruction and guidance. None of the health/medical information in this book is intended as a substitute for professional medical advice; always seek the advice of a qualified practitioner.